The NASA Letters

By

Don A. Nelson
Retired NASA Aerospace Engineer

Second Release 8/2017

ISBN: 1542763886
ISBN 13: 9781542763882
Library of Congress Control Number: 2017901240

Contents

When we fail to learn from our mistakes we will most certainly fail again.

Preface

The landing of humans on our Moon was thought to be the first step in taking mankind to the planet Mars and beyond. Instead, nearly half a century later we cannot get our astronauts to the space station without the consent of a foreign government. The return to the Moon program has been canceled and the plans for a human Mars mission are fading. As a NASA aerospace engineer (now retired) it is my professional judgment that the reason this nation's human space flight programs have failed is because NASA's senior human space flight management has become incompetent and is not being held accountable. While this may seem like a harsh indictment, what the reader will learn from the disclosures in *The NASA Letters* is that NASA's human space flight management has:

- Misrepresented the safety of the Orion and commercial space crew modules to be used for ferrying astronauts to and from the space station and for deep space missions. Crew modules are one failure away from being death traps.

- Distorted the capabilities of the Ares V lunar launcher and the Space Launch System's launchers to deep space and Mars by repeated unsubstantiated claims they will be safe, affordable, and sustainable. In reality the Ares V launcher has been canceled for failing to achieve any of these attributes, and there is no data to support that the Space Launch System's vehicles can meet these claims.

- Failed to provide this nation with an affordable launch system that can compete in the international launch market and provide a rapid response to military and deep space threats.

One of my NASA Apollo managers had a plaque on his desk that read, "Don't bring me any problem without also bringing me a solution." The intention of *"The NASA Letters"* is to make the reader aware that NASA's Space Shuttle replacement vehicles and deep space launchers have unsolvable problems. The solution is a Commercial Space Shuttle (CSS) freighter which is the first step on the only path forward to a 21st century affordable and safe human space transportation system. NASA's human space management has steadfastly rejected the CSS freighter.

Note: Names have been removed from the enclosed letters and emails except for those cases where the sender's name is obvious. The objective of this book is not to blame anyone for the deplorable status of our human space endeavors but only to identify the reasons for the problems and provide a solution.

Emphasis is added by printing important information in bold type.

Introduction

It was just a small blip in the star filled view of his homemade telescope when the backyard astronomer first noticed that it was something he hadn't seen before. Because of bad weather he didn't get back to take another look until three days later. To his surprise, the blip was much larger. Hoping he might be the first to report this unknown object he immediately notified the Amateur Astronomers Association. It was quickly determined that the amateur astronomer had found an asteroid; and, instead of being named for him, it was given the name *"Bad Day"*. An unusual name for an asteroid, but this one merited the name because NASA's Near Earth Asteroid Tracking group calculated that asteroid *Bad Day* was on a keyhole trajectory to impact Earth. It was not a doomsday asteroid but still would do major damage when it impacted Earth. Astrophysical scientists needed to know the composition of the asteroid before a plan to stop it from impacting earth could be devised. A nuclear explosion might deflect a solid iron core asteroid but could shatter one composed of other materials and expand the impact zone. Time was crucial because impact was calculated to occur in the northern hemisphere in two months. A survey of what launch vehicles were available to send probes to the asteroid found that none could meet this dire schedule for launch. Asteroid *Bad Day* would impact our planet Earth and there was nothing we could do to stop it! Thousands of people could be expected to perish and the recovery could take decades.

Okay, so maybe the above scenario is fear-mongering, but it does point out the consequence of not having a rapid response launch system. The same scenario would be the case for a military threat or cosmic storm disruption on our satellites. We must have an affordable rapid response launch system. Developing and operating an affordable rapid response launch system can be achieved with existing technology. First, the launcher must be reusable. Second, the government cannot be the operator of the launch vehicle(s). Third, being reusable and removing the government management makes the launch system competitive in the commercial launch market. Income from the commercial market is mandatory to provide sustainability by eliminating the uncertainty of government support. For years we've known these requirements, but the one attempt by NASA to develop a commercial reusable launch system was mismanaged by NASA. Other attempts by myself and others to privatize the Space Shuttle or use existing technology to develop a commercial Space Shuttle or a Space Shuttle freighter were met with steadfast resistance from NASA's human space flight management. What the reader will discover is there are no technical issues that would prevent the development of a commercial Space Shuttle freighter. This old NASA engineer and other Concerned American Aerospace Engineers (CAAE--a group of supporters and information contributors) are convinced that a commercial Space Shuttle freighter can be developed and operated by the private sector of the aerospace industry, which will be superior to the expendable launchers and crew modules that NASA is struggling to develop. This is not a technical problem. The technology existed which could have been used to develop a commercial space shuttle transportation system. What we have is a cultural roadblock kept in place by senior NASA human space flight management who steadfastly has resisted the efforts to allow the private sector to develop and operate human space flight vehicles. This obsolete cultural bias has this nation's space transportation capability on the verge of collapse and again--in this old NASA engineer's opinion--left mankind exposed to the threats from deep space.

The readers are requested to evaluate the letters and emails in this book and decide for themselves if NASA's human space flight management and their oversight panels have exercised due diligence in the management of the human space flight transportation systems. Has NASA made every "reasonable" attempt to provide and/or insure our astronauts will have the safest space transportation possible? Did NASA space flight management conduct any "reasonable" evaluation(s) that proved expendable launch systems would provide the U.S. Taxpayer a more sustainable and affordable human space transportation system than the Space Shuttle, a privatized Space Shuttle and/or an automated commercial Space Shuttle freighter?

Let me issue you a challenge. After reading *The NASA Letters* if you think our human space program is being mismanaged, speak out to the President and Congress and media. OUR VERY SURVIVAL MAY HINGE ON THIS ISSUE...DON'T REMAIN SILENT!

Note: This old NASA engineer really believes there is a "Bad Day" asteroid/comet out there and that it's just a matter of time when...

Chapter 1 Why was Apollo Canceled?

Apollo 15 NASA

There have been lots of books written on the success of the Apollo program. I won't dwell on the details of why the program was so successful except that it was my observation that the goal was well defined and we had exceptional management. By using the existing military missile launchers, the Atlas and Titan II, for launching the Mercury and Gemini manned spacecraft, we were able to verify the human capability to function in space and establish the trajectory techniques for rendezvousing space vehicles. While these unproven feats were being accomplished, some of the largest launch vehicles ever built, the Saturn I and Saturn V, were being developed and flight tested. Not enough can be said about the Apollo program management teams and their achievements. So what went wrong? Why are humans still confined to low earth operations?

Space Race

I went to work for NASA in Houston, Texas in 1963. I was one of those twenty year olds lucky enough to be a part of the goal to land men on the Moon...the Apollo program. Unlike many of my twenty year old colleagues I actually had work experience with space vehicles, having done launch performance analyses on the USAF Titan I ballistic missile and the NASA Scout launch vehicles. The Titan launcher was to be used to launch the Gemini spacecraft, and I was assigned to help develop a computer program to simulate the flight mechanics (trajectory) of the launch vehicle ascent to orbit. For the program design I needed to know the mass properties of the vehicles: the weights and moments of inertia. I was told that an old guy from the Langley center had that data. Also that he was on temporary assignment here in Houston, was getting ready to retire, and that he was one grouchy old guy who gave the twenty year olds a hard time. So it was with some reluctance that I went to see him for my needed data. Sure enough, he started to berate me before I even asked for the data. But his mood changed when I explained what data I needed and why I needed it. What he realized was that I could talk the talk and knew what I was doing. After he got my data we talked; he was interested in the launchers I had previously worked on. Then he told me something I've never forgotten: "Son, if we were at the Langley center you would have needed a work order with five signatures on it before I could have given you this data. The bureaucracy is not yet here in Houston, enjoy it while you can, but it will be here someday. What needs to be done is every twenty years they should shut down these NASA centers and moved them fifty miles down the road to get rid of the deadwood people and the ingrained bad cultural." The Johnson Space Center has been there over fifty years.

Apollo Operations

At the peak of program over 400,000 people were employed on Apollo, which by its very size should have been a management nightmare. It wasn't a nightmare because we had two management organizations. One was the "official" on paper organization which really only served as the management of personnel matters and the other was the working panels where the actual management of the program was conducted. We had one team working from many different organizations, including contractors. My work assignment was with mission planning and flight operations. My panel was designated as the Flight Operation Panel. Panel members came from different organizations and represented different engineering driplines such as aerodynamics, mass properties, rocket propulsion, or guidance. My work assignments came from the panel manager, and as we neared flight time the work assignment changed from defining the flight capability of the flight system to using that gained information to plan the flight. If we had tried to manage the Apollo program within the structure of the official organization charts it would have been untenable.

I, like most of my colleagues, assumed that the Apollo program would continue after achieving the goal of beating the Soviets to the Moon. What we hadn't considered was the operational cost of the Apollo lunar missions. From 1963 to the program cancellation in 1972, the Apollo budget was an average of over 50 percent of the total NASA budget. In 1967 it was 70 percent of the total budget. The 1970 budget required 54 percent of the total NASA budget to conduct an average of two Apollo lunar missions per year.

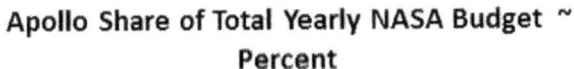

Apollo Share of Total Yearly NASA Budget ~ Percent

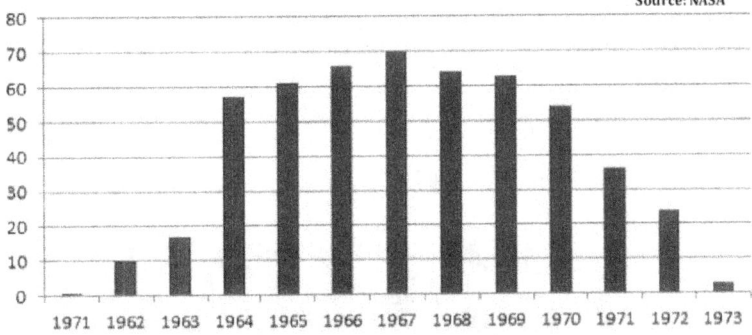

The science community attacked the Apollo missions as having limited science results which didn't justify the cost. The Apollo crew module could only return 400 pounds of material from the Moon and no significant minerals with current commercial value had been found. The military had no missions for Saturn V launchers, and there were no commercial payloads that required the heavy lift launcher. Final analysis was that the Apollo program had become an unaffordable and nonproductive program. We needed to get the launch cost down to an affordable level (still not there after more than 50 years).

Note: The operations cost of the Apollo program in 1970 was $12 billion (FY 2017 dollars) for two missions or 67 percent of NASA's total 2017 budget. While it may be feasible for human missions to the Moon and Mars with Saturn class launchers, it is definitely not realistic in today's economy.

Apollo Safety Concerns Ignored Today

We were aware that we could lose astronauts in our quest for beating the Soviet Union to the Moon. So when the fire in the Apollo 1 command module took the lives of the crew, it was a horrible shock but also a fear realized. Just the week before the fire, I had a phone conversation with an Apollo 1 crew member about a contingency plan for deorbiting if the service module engine failed to function for the deorbit burn. I was working on a technique to use the small attitude control engines in the command module to deorbit the crew, but just couldn't find a way to verify the process in our simulation program. So I called my crew contact and he told me they had been able to simulate the deorbit maneuver that morning in the crew trainer. So we had a solution for another "what if" problem. Somehow we missed the "what if" electrical short circuit that caused the fatal Apollo 1 fire, and we paid a terrible price for that error.

Aftermath of Apollo 1 Command Module Fire (NASA)

The electrical short caused a flash fire, and with only one access hatch the crew didn't have time to get out. Getting the crew out of the command module in case of an emergency is one of the many reasons command modules are death traps. The small compact configuration of command modules prohibits installing any type of escape system. As one of my engineering colleagues at NASA told me, **"In command modules all critical systems work or the crew dies."** So why now, in 2017, do we put astronauts in the Russian Soyuz crew modules, and all our future crew carrying spacecraft will be crew modules?

There are other Apollo safety problems that NASA management has ignored in their plan to return to the Apollo class transportation systems of heavy lift launchers and crew modules. If the Apollo 13 hydrogen tank explosion had occurred on the Apollo 8, where there would have been no Lunar Lander spacecraft to make the return to Earth maneuvers, the crew would have been lost. If the one parachute-failure-to-open on Apollo 15 had occurred on a rough sea state condition day, the crew would have been lost or severally injured.

Apollo 15 Parachute Failure (NASA)

While safety is now being ballyhooed by having launch escape rockets to get the crew modules away from an exploding launcher, there is no system that can protect the crew while in orbit or during entry. An event causing a decompression of the command module can and has killed a crew. The only way to insure crew safety is to have crew escape pods, but that is structurally impossible to install on crew modules.

There is another safety hazard that cannot be avoided for any expendable launcher like the Saturn V, and that is manufacturing error(s). The parachute failure on Apollo 15 was caused by a manufacturing error. Every launch of an expendable launch vehicle is the maiden flight of flight systems that have never been flown. While preflight testing has done a remarkable job verifying that a subsystem is flight qualified, it is the real world environment of launch and entry vibrations and heating that is the final test the system must pass before it can be declared safe. In those rare but real cases it is the time disaster strikes. To continue with expendable launchers only continues the path to inevitable launch failure (further explanation later in book).

Summary of Apollo lessons not learned:

1) The Saturn V launch system was too expensive to continue because of the manufacturing and operation cost.

2) Command modules for crews are inherently unsafe because they can't incorporate crew escape pods and cannot return a cargo mass sufficient for commercial operation because of the weight limitation that parachutes can carry.

3) There is always the possibility that a catastrophic manufacturing error will not be detected until after the launch of an expendable launcher.

Flight Ready Saturn V in Johnson center "Mausoleum"

After the Apollo program was canceled there were three completed, ready to launch Saturn V launchers. One was used to launch the Skylab space station and the others are in museums (mausoleums) at the Kennedy and Johnson Centers. Both are entombed because this nation couldn't afford them then and we certainly can't afford these obsolete massive launchers now. The Space Shuttle proved, with the construction of the space station, that assembly in orbit is the far better way to build our space infrastructures. So why is NASA management returning to the failed Apollo/Saturn class launch systems?

Chapter 2   The Space Shuttle's Bad Deal

"The Space Shuttle got a bad deal." **NASA Astronaut**

The Space Shuttle was first publicly presented in the 1969 Report of the Space Task Group. The report stated:

"Exploration and exploitation of space **is costly with our current generation of expendable launch vehicles and spacecraft systems.** *This is particularly true for the manned flight program. Recovery and launch costs will become an even more significant factor when multiple re-visit and resupply missions to Earth orbiting space station are contemplated. Future developments should emphasize:* **A reusable chemically fueled shuttle operating between the surface of the Earth and low-earth orbit in an airline-type mode."**

The goal was to lower the cost of launch operations by developing a "reusable" space shuttle. The Space Task Group report acknowledged that expendable launch vehicles were too costly especially for manned flight programs. However, now in 2017, we're back to trying to develop an expendable manned launch system. What happened was the Space Shuttle got a bad deal.

My introduction to the Space Shuttle concept was at a meeting in the Building 30 auditorium at the Houston Manned Spacecraft Center (now Johnson) in 1972. Needless to say the auditorium was packed, as we were anxious to learn what our jobs would be after Apollo. As I listened to the speaker I became very doubtful that this shuttle thing would ever get off the ground. Just building an aircraft that could go to orbit and return to land was mind boggling, but the speaker was claiming that it would be able to launch once a week for a cost of $15 million (1972 dollars). I wasn't the only one shaking my head in disbelief that day. But hey, if our management said it could be built, then we would build it. After all, we did just put men on the Moon!

There had been a number of iterations on the design of the shuttle while we were working Apollo. The design chosen was called a stage and a half concept because only the spacecraft stage called the orbiter would be reusable. Two large solid rocket boosters and a large external tank carrying propellants for the three reusable engines on orbiter would be expendables.

Not the best configuration but one that could be supported by existing technology or technology that was thought achievable. All we had to do was develop a reusable engine that could produce at least 400,000 pounds of thrust and a thermal protection system that could protect the orbiter structure from 2000 degree Fahrenheit entry temperatures. And also make the engine able to increase propulsion up to 109 percent of the rated thrust, and keep the thermal protection tiles as light as possible, because the 150,000 pound design dry weight of the orbiter had been exceeded by 30,000 pounds. Oh yes, we were told to make the orbiter safe enough that the crew could fly in a commercial airliner shirt sleeve environment and not need any type of escape system. Another thing, if you're not too busy let's use up the leftover Apollo spaceflight hardware by building and operating the Skylab space station and conduct a goodwill Apollo spacecraft rendezvous with the Soviet Soyuz. Sorry the money is a little tight so do this as cheap as possible.

So, from the end of Apollo in 1972 till the first flight of the Space Shuttle in 1981, we in the manned space flight business were a bit busy. Where the Apollo program had been an achievement for mankind, the Space Shuttle program was an engineering marvel. I was at the Paris air show in 1983 watching the shuttle Enterprise fly-by when I heard the Frenchman standing next to me say, "magnificent, magnificent, magnificent!" There is no question that the Space Shuttle was a magnificent flying machine. In 135 missions; seven payloads were retrieved, repaired, and re-deployed back in orbit. It carried eight astronauts on one mission, stayed on orbit for over seventeen days, and deployed payloads weighing twenty-five metric tons. Many payloads like the Hubble space telescope are still functioning today.

So what went wrong with the Space Shuttle program? Some have said that the program was underfunded, and there is some truth to that. Others point out that technologies needed to have affordable launch operations were not available, and there is some truth in that. However, what really happened was we mismanaged the operations of the Space Shuttle program. I say "we" because I was a NASA Space Shuttle engineer and manager and I speak from that failed experience. We really blew it and we continue to mismanage our human space program endeavors. Today there are existing technologies that will make the reusable shuttle concept a safe and affordable launch system (details in later chapters), but today the NASA mismanagement also still prevails, so we're still stuck in low earth orbit.

The shuttle's reusable orbiter stage was to be a space truck capable of launching one hundred times a year. Instead, the maximum number of Shuttle launches in one year was nine, and a University of Colorado study showed the average cost per launch to be over $1.5 billion. With the loss of two Space Shuttle crews, the launcher was declared unsafe and unaffordable and was decommissioned in 2011.

However, we got off to a very good start and had twenty-four successful missions before we lost the Challenger shuttle. The good start made us overconfident and in the space launch business that is always a fatal mistake. After four missions we removed the two ejection seats. The seats were only available for the two pilots and were only considered safe to use below 10,000 feet. At that time we were flying mission payload specialist astronauts, and the shuttle commanders refused to leave them to a certain death if the pilots ejected. In effect, this decision declared that we were no longer in a flight test program and the Space Shuttle was now an operational launch system. There was no official announcement that the vehicle was operational because after every flight we had anomalies that had to be addressed before the next flight. We were still a long way from the goal of an airline-type operation that provided safe and reliable transportation, but we were operating the shuttle like it was a flight safety proven launcher...it wasn't.

Challenger Lost January 28, 1986

I always liked to get to work early, plan my day, read mail and reports, or just take time to think without any phone call interruptions or someone dropping in to talk. However, this fateful morning was different; the Building 4 parking lot was already nearly full when I pulled in to park. It was launch day for the Space Shuttle Challenger, the "teacher in space" mission. The flight controllers assigned to the launch team were already at their consoles in the mission control center.

When I got to my office I turned on my mission control center communication loop speaker to check the status of the mission. There was some chatter from a supporting backroom flight controller, and the front room primary flight controller was talking about the weather at the launch center. Then the flight director cautioned that it was time to get serious and the com loop went quiet. The Kennedy Center launch controllers would be in charge until liftoff, and then the Johnson Center controllers would be in charge.

For the round-the-clock flight control operations there were three flight control teams, each having around 75 engineers to monitor the mission. Shown below are the "front room" flight controllers at their consoles monitoring different shuttle flight systems. They all had worked hard in preparing for these prestigious positions. One of the flight directors had labeled them the "glory hogs" because they are the ones seen on TV during mission coverage.

Mission Control Front Room (NASA)

Each controller had a backup support team in the "backrooms" of the mission control center. The backroom controllers could call on off-duty systems engineers if the need occurred and time permitted. The three teams of controllers each had their own flight director and each one worked an eight-hour shift. One team would specialize in launch and another entry. These teams had spent months training for this mission and they were more than ready to get this show on the road. The following is the NASA launch timeline post flight account and my recollections of the Challenger launch:

T minus 2:00:00 (hours, minutes, seconds to liftoff)

The crew was having breakfast and the first of three inspections of the icing conditions at the launch pad was made.

From years of waiting for a mission liftoff I never got enthused about the actual launch until the countdown got to 4 minutes, and even then something could cause the mission to be delayed or scrubbed. This mission had been scrubbed the day before because of high winds and today the weather at the Kennedy launch site was again an unknown factor. With the mission control communications quiet I could start my day.

T minus 1:00:00

A voice on the com loop informed the flight director that the crew was strapped in their seats, but the Kennedy launch director had declared a launch hold. They were still concerned about an accumulation of ice that had built up on the launch pad structure. There would be a two hour hold to let the sun melt the ice on the pad and avoid any ice damage to the orbiter heat shield tiles.

Two Hours and Counting

At 10:25 AM Houston time someone on the com loop announced that the Challenger count was starting. The crew had been strapped in their seats for nearly three hours. There had been three inspections for ice. The temperature measure on the right solid rocket booster was 19 degrees Fahrenheit, 14 degrees colder than the left booster. Management had decided the ice was not a problem and given the mission a go.

T minus 10 Hold

This was a planned hold to switch from launch pad ground power to internal shuttle power. For me it was a signal to gather in the building 4 lobby with others not assigned to this mission to watch the launch on a TV monitor. There were fifty or more of us that morning in the lobby waiting for the launch countdown to restart. The overhead TV monitor showed the Challenger sitting on the launch pad. Most of the conversations in the lobby, were about the Chicago Bears' trouncing of the New England Patriots 46-10 in Super Bowl XX. Two flight controllers were in a heated discussion on the mission's STS 51 L number designation. Both were wrong in what they thought the designations meant. The Space Shuttle is officially referred to as the Space Transportation System (STS). Initially, the launches were given sequential numbers indicating order of launch, which we used for STS 1 thru STS 9. Beginning in 1984, each mission was assigned a code with the first digit indicating the last digit of the federal fiscal year. The second digit was for the launch site. Kennedy Space Center was 1 and Vandenberg Air Force Base in California was 2, and the letter indicating scheduling sequence. Both flight controllers agreed that only someone at NASA headquarters could have come up with something this confusing.

Another group was discussing this year's chili cook-off. Teams with chili names like Kangaroo, Outhouse, and Road Kill didn't make for a gourmet delight. Before the contest, cook-off teams spent two weeks advertising the merits of their chili with signs posted everywhere in the building. This year they went too far when they put Playboy centerfolds in the ladies restrooms with suggestions that their chili was sexually enhancing. Someone complained and we got a lesson in the new sensitivity rules. One of the teams somehow got a VW Thing convertible car decorated with their chili logo into the lobby of the building. Center security went into gimbal lock (rocket term meaning loss of control), but backed off when they couldn't explain how the "Thing" got past their guards.

T minus 10

The launch countdown is picked up and the clock starts.

There was only a slight pause in the lobby conversations with the announcement of the restarting of the countdown. Having experienced so many holds and launch cancellations these veteran controllers knew not to take the count at this time too seriously.

T minus 8 minutes and counting

All flight recorders are on.

I overheard one flight controller talking about their assignment on the next mission, which was to put the Hubble space telescope in orbit. He was excited about being part of a mission that held the promise of unlocking some of the secrets of our universe. The Challenger mission, with the exception of the teacher in space aspect, was now considered routine. The primary goal mission was to place a needed NASA communication satellite in orbit, but the upcoming Hubble telescope mission was definitely exciting.

T minus 7

The crew access arm is retracting pulling away from orbiter.

This was the tenth flight for Challenger. It was the twenty-fifth for the Space Shuttle program. It was the start of a very busy year with nine flights scheduled. After the "Teacher in Space" flight we would next fly a journalist in what the White House called the "bring people to space" program. With twenty-five successful flights the shuttle was thought to be operational and safe. However, there were warning signs.

T minus 4

The crew is reminded to shut the airtight visors on their helmets

T minus 1:46

The so-called beanie cap device designed to keep ice from forming on the external tank is removed and retracted.

T minus 0.00

All systems are go and we have a liftoff.

Lobby voices: aw right... yesss...we're go. Then there was quiet as everyone listened for the next call.

KSC com.: "liftoff of the 25th space shuttle mission and it has cleared the tower." Launch control shifted to Mission Operation Center at Johnson.

T plus 1:00

Shuttle pilot intercom: "Here we go."

As a private pilot, I understood what the shuttle pilot was feeling. That first movement of flight speaks to your mind with signals of fright and elation. It's also the feeling that one gets when you're a passenger in an airliner and the engines roar and you start down the runway.

There is a saying about flying: "It's sometimes hours of boredom with thirty seconds of sheer terror." With a shuttle launch it is never boring.

At T plus 0.678 was the first sign that the Challenger and its crew were doomed. *Cameras at the launch pad record black smoke swirling out between the right hand solid rocket booster and the external tank.* The flight controllers are unaware of what has happened.

The Launch commentator on NASA-SELECT television announces: "*Liftoff of the 25th space shuttle mission, and it has cleared the tower.*" Someone in the lobby says, "Go big bird...go."

T plus 27:000

Booster systems engineer (Booster) mission control: "Throttle down to 94."

Challenger's three main engines begin throttling down as planned as the launcher approaches the region of maximum aerodynamic pressure.

T Plus 45.000

Launch commentator: "Engines are at 65 percent. Three engines running normally..."

T plus 54:000

Launch commentator: "Velocity 2,257 feet per second (1,539 mph), altitude 4.3 nautical miles,

T plus 58.788

Tracking cameras show the first evidence of an abnormal plume on the right-hand solid rocket booster facing away from the shuttle. Crew had no data on the performance of the solid rockets except for a software system that would have alerted them to malfunctions in the booster steering mechanism.

T plus 59.262

A plume of exhaust is seen on the side of the right booster by tracking cameras. This is clear evidence of an O-ring joint burn through.

T Plus 66.000

Booster systems engineer: "Throttle up, three at 104."

The three main engines, thrust level was advanced to maximum thrust level with each engine producing over 400,000 pounds of thrust.

T plus 73.191

A sudden brilliant flash is photographed between the shuttle and the external tank. TV tracking camera: Fireballs merge into bright yellow and red mass of flame that engulfs Challenger.

One of the flight controllers watching the launch in the lobby spots the plume and yells out, "Where's that light coming from?" Before anyone could reply there was a fireball and the Challenger came apart. Stunned voices filled the lobby: "Oh No!...Oh my God, oh my God...It's gone, We've lost the crew!"

The lobby became silent. Just seventy three seconds after launch and nine miles high, the Challenger was just flying debris. My first thought was, "We let the teacher down."

T plus 73.213

An explosion occurs near the forward part of the tank where the solid rocket boosters attach.

The Challenger had been at 46,000 feet and traveling nearly twice the speed of sound, and now we watched in silence as pieces of the vehicle went spiraling off in different directions.

T Plus 89.000

Ground control engineer mission control: "Flight, we've had negative contact, loss of downlink (of radio voice or data from Challenger).   Flight Director: "OK, all operators, watch your data carefully."

The flight controllers were not getting any information from the Challenger's communication systems. The realization that the Challenger had been lost was becoming evident.

T Plus 110.250

Range safety control officer sends radio signals that detonated the self-destruct package on right-hand solid rocket.

T Plus 110.252

The left-hand booster self destructs.

Range safety had to send the destruct signal because their tracking data indicated that one of the boosters was heading back towards land. The range safety officer that sent the destruct signal knew that now there was no hope for the crew. Their duty now was to prevent any more loss of lives.

T Plus 2 minutes 25 seconds

Flight director asked for any reports from recovery forces.

There is the silence of death in the lobby. Everyone knows the recovery forces will only be able to recover debris and bodies of the crew. Knowing looks were exchanged and we drifted slowly back to our offices.

T Plus 6 minutes 39 seconds

Flight Director asked everybody to stay off the telephones and make sure you maintain your data...start pulling it together.

After years of planning and training, all the Flight Director could do was issue the command to **"secure your data!"**

T Plus 15 minutes 6 seconds

There is a report that no recovery forces have been deployed. Debris is still coming down in the recovery area.

Back in my office and I turned on my speaker phone to the flight controller com loop. The Flight Director asked how long it would be before the rescue vehicles would arrive at the impact site? He was told it would take fifty minutes for the debris to fall and then the recover teams could go into the impact area.

T Plus 21 minutes 53 seconds

Launch commentator from Mission Control: "This is mission control, Houston. Repeating the information that we have at this time. We had an apparently nominal liftoff this morning at 11:38 Eastern time. The ascent phase appeared normal through approximately the completion of the roll program and throttle down and engine throttle back to 104 percent. At that point, we had an apparent explosion.

Challenger Lessons Not Learned

The accident investigation would conclude that a seal in the solid rocket booster had failed because of the cold weather prior to launch. Hot gases bypassed the ruptured seal and sprayed onto the external tank causing it to also rupture and explode. There were engineers who had warned that the booster seals could possibly fail in a cold launch environment, but those warnings were ignored. As the reader will learn in later chapters, NASA human space flight managers still reject warnings that conflict with their agenda.

At the time before the flight controllers lost contact with the Challenger they had no indication of any problems. There was an important lesson here but we missed it. In space flight operations something can go wrong in a millisecond and you may have only another millisecond to take action. Waiting for a pilot or a flight controller to take corrective action will in all likelihood be too late. Only an automated flight control system can react in milliseconds. With automated flight control you don't need a large number of engineers to monitor the mission.

I didn't learn of the most important finding until some years later. At the time when the crew compartment impacted the water, some of the crew was believed to be still alive. Had there been some type of escape system it is possible they would have survived. Our goal for a safe airline environment had killed the crew. We had investigated crew escape systems but the weight of the systems made the orbiter nose heavy causing the vehicle to stall during entry. There was a solution for this problem, but again, I didn't learn of this until many years later (solution discussed in later chapter).

Before the Challenger disaster, NASA had been launching commercial payloads and the manufacturers of expendable launchers had complained to President Reagan that using a government vehicle to launch commercial payloads was unfair. Their complaint was justified, but only the Space Shuttle had the unique capability to return the payload if it didn't work or needed on-orbit service. After Challenger in 1988, Reagan issued a "Presidential Directive on National Space Policy," which stopped commercial payloads from being flown on the Space Shuttle. Responsibility was to remain in NASA for operational control of the Space Shuttle for civil missions. Prior to the Challenger disaster, NASA had 44 commercial payloads planned for shuttle missions. With the loss of commercial payloads there would not be enough launches to keep the operation work force busy. The result was a dramatic increase in launch operations cost. To keep the launch cost down, the Space Shuttle operations should have been transferred to the private sector. In my professional opinion this failure to privatize operations was the beginning of the demise of the Space Shuttle.

Challenger Investigation

The 1986 "Report of the Presidential Commission on the Space Shuttle Challenger Accident" (also known as the Rogers Commission report) was very critical of the management process that permitted the Challenger launch in such cold conditions. Morton Thiokol engineers had made known their safety concerns about a failure of the O-rings in cold weather; however, the commission concluded that the management oversight was inadequate and led to the fatal decision to launch. One of the commission's recommendations was to: "*Make all efforts to provide a crew escape system for use during controlled gliding flight."* At the time of this report it was believed that the Space Shuttle could not accommodate the weight required for a crew escape system. However, it is now known that a shuttle crew escape system is feasible.

A disturbing, and I believe still valid, NASA safety failure was the observation of Dr. Richard P. Feynman, theoretical physicist and 1965 Nobel Prize in Physics recipient. Dr. Feynman stated in the commission report: *"Finally, if we are to replace standard numerical probability usage with engineering judgment, why do we find such an enormous disparity between the management estimate and the judgment of the engineers? It would appear that, for whatever purpose, be it for internal or external consumption, the management of NASA exaggerates the reliability of its product, to the point of fantasy."*

In my professional engineering judgment the management fantasy reliability problem still exists at NASA, and the reader will learn in later chapters that NASA still promotes crew safety with false numerical probability analyses.

Columbia Lakes

It would take nearly three years to get the Space Shuttle back to flight status. We used the time to check all our mission rules and procedures. All subsystems on the orbiter were reevaluated and documentation was updated. Computer programs used for flight operations were reviewed and documented. We took very seriously the task to make the Space Shuttle a safe human transportation system. As we finished up our review, we began to have time to take some professional training courses. These type courses were popular at the time, and a number of companies had developed a curriculum that was thought to provide in-depth problem solving training for managers. I was chosen to attend one at the Columbia Lakes resort located about 50 miles south of the Johnson center. The course lasted two weeks and we stayed the whole time at the resort. We were separated into small groups and assigned a roommate. There was only one other mission control engineer in my group of ten. All the others were from the program office, engineering, accounting, and even a medical doctor. This provided a good mix of backgrounds for the problems solving exercises. The teachers really knew their material, and without the interference of any outside activities we soon bonded as a team. Senior managers came down to the resort and gave their thoughts on the future of the Center and the status of returning to flight. Even though our daily problem solving exercises had nothing to do with the Space Shuttle, it was the only thing we had in common and as a result most of the off-time conversations were concerning our jobs. From the management briefings and these conversations it became evident that the management solution for the Challenger disaster was to put in place more personnel and procedures to verify that all Shuttle preflight procedures had been properly checked before launch.

This was to be another disaster. When we got back to flight planning we became bogged down in paperwork. Many in my Columbia Lakes group had warned their management that we didn't need more people, but were ignored. In my flight planning I had a form that required over twenty signatures for approval before flight. Only three people knew what they were signing. By 1993 the Space Shuttle operations work force exceeded 21,000 employees. More people resulted in more problems and an increase in launch costs. We were in a countdown to another shuttle disaster before we even got back to launch status.

Advance Launch Programs...*It's the launch cost...stupid!*

If there was any positive outcome from the Challenger disaster, it was that NASA management started reviews of advanced launch systems to replace the Space Shuttle fleet. What was a surprise was how strongly and sometimes deviously NASA management would oppose the only feasible and realistic option to reduce operation cost and improve safety, which was to automate the shuttle flight operations and privatize the shuttle fleet. It is obvious from the following chart that operation costs are the key contributor to launch cost, and only a Commercial Space Shuttle (CSS) can significantly reduce launch costs.

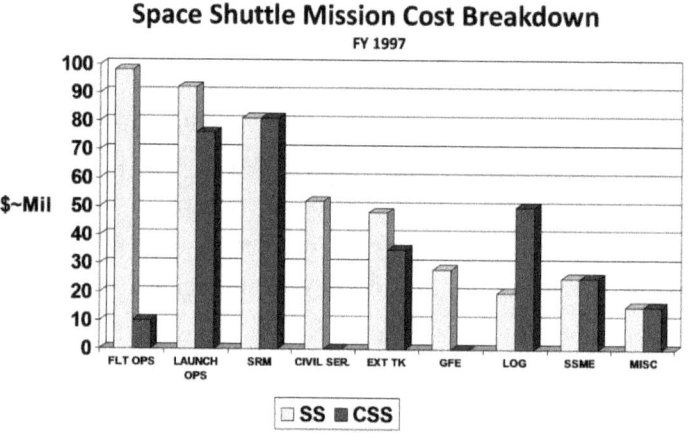

The CSS launch cost of $205 million per mission was for seven launches per year using NASA operation procedures and data. Competitive launch cost of less than $100 million per mission can be obtained by increased number of launches, reduced solid motor and composite external tank cost, and integrated flight operations. Additional launch cost reduction and increased reliability can be achieved by fly-back boosters (discussed in later chapters). NASA management has challenged the validity of these cost numbers for the Space Shuttle (SS) and Commercial Space Shuttle (CSS), and their challenge is indeed valid. The cost numbers have been gathered from numerous sources and in some cases are educated guesses. Also, NASA's shuttle cost accounting was dubious. For example when I worked advanced launch systems, I had to charge my time to shuttle operations. However, the reader will learn in the following chapters that NASA management has steadfastly refused to conduct evaluations on the cost for an automated flight system and privatized shuttle fleet operations. In 1991, NASA was forced by a 107th Congress request to investigate privatizing the space shuttle program (SSP). A summary from the reports follows:

CONCEPT OF PRIVATIZATION OF THE SPACE SHUTTLE PROGRAM
September 28, 2001
XXXX XXXXXX
Manager, Space Shuttle Program
Summary Excerpts:
It is believed that utilization of the Space Shuttle for human access to space will continue through at least 2015 and possibly beyond 2020. ***The longevity and operational aspects of this program demand a different approach to operational management for the future. A different management strategy needs to be employed.***
Privatization of the SSP has the potential to provide significant benefits to the Government. However, timing is critical. *The continuing erosion of NASA skills and experience threatens the safety of the program. It is critical to take advantage of the existing NASA SSP expertise before further erosion affects the ability to plan and safely implement privatization. Today, the skill and knowledge legacy still remain to formulate the appropriate merger of the NASA SSP and private industry.*
From a business point of view, the focus of the private company's marketing program will be the full use of the Space Shuttle for the benefit of the nation. ***The private company will be free to develop customers for the Space Shuttle.*** *These customers will include NASA as the anchor tenant along with other United States (U.S) government agencies, other space faring nations, and U.S. industry.*
Ref.: http://www.hq.nasa.gov/office/hqlibrary/documents/o56403525.pdf

In the report, NASA acknowledged that the Space Shuttle's operational management needed to be changed and that privatization had the potential to provide "significant benefits." Privatization allowed commercial payloads to again be carried on the shuttle which had the potential to reduce launch cost by half by the resulting increase in number of launches. This meant the aerospace companies building expendable launch vehicles would lose customers and that business would collapse, a reason the shuttle fleet was never privatized...but there is a far more disturbing factor.

This report also failed to consider automation which improved safety and further reduced operations cost. The shuttle was already nearly automated with only two functions requiring a pilot's action -- deploying the wheels in the down position for landing and releasing the drag chute to slow the runway landing speed. Automation improved safety by removing potential pilot errors in time-critical events like a launch abort back to the landing site. Automation would also make possible an extremely important crew safety mechanism...installations of crew escape pods. A crew escape system was never installed because its weight in the forward section of the orbiter made the vehicle nose heavy and aerodynamically unstable during atmospheric entry causing a stall and a crash. However, the problem would have been solved with automation because crew escape pods could be installed in place of the removed weight of two pilots, windows, and crew displays. **IF THE SPACE SHUTTLE COLUMBIA HAD CREW ESCAPE PODS IT IS HIGHLY PROBABLE THE CREW WOULD HAVE SURVIVED!**

Summary of Space Shuttle Lessons not learned:

1. The Space Shuttle should have been automated. Most shuttle missions could have flown without a crew, which would have reduced operations cost and not exposed the crew to unnecessary risks. Crew safety would have been increased by having the weight margin needed to install crew escape pods. A primary reason automation was not approved was that NASA management failed to acknowledge that sensor technology used to monitor the status of the vehicle was significantly better than that used in the Mercury, Gemini, and Apollo programs.

2. The Space Shuttle flight operation should have been privatized. This failure by management insured that launch costs would be

unacceptable. Launch costs are a function of flight rate, and with President Reagan's directive prohibiting commercial payloads to be flown on the shuttle, the launch cost soared. There was no valid reason not to privatize the Space Shuttle operations.

Chapter 3   NASA's Advance Launch Systems Fails

The Challenger disaster would send the nation's space transportation endeavors into one failure after another and put the shuttle fleet in museums. What follows in this and later chapters are the obstacles this engineer encountered in getting NASA to automate and privatize the shuttle fleet.

Career Ending Assignment...Advanced Launch Systems

I had worked the first 30 shuttle missions and the job had become a "turn the crank" operation. As an engineer I wanted something more challenging. I thought that investigating advance launch vehicles concepts would be very challenging. Before coming to NASA I had worked as a contractor doing launch performance and trajectory calculations on the Air Force Titan I ballistic missiles and the NASA Scout launch vehicle. My shuttle assignment conducting consumable requirements (propellants, water, electrical power) for the orbiter gave me a good background in flight systems as they related to the mission requirements. So when an opening in 1989 was announced for the position of office manager of advanced launch systems in the Mission Operation Directorate I applied. I didn't get the job but was one of the two engineers assigned to the office. Little did I realize that this challenging work assignment would eventually end my NASA career.

For the next 11 years until my retirement in 1999 I would participate in all the NASA advanced launch system studies. I've lost track of how many studies we conducted. Some were unfunded and other ones were funded, and some were well publicized, such as the Space Exploration Initiative, Space Launch Initiative, First Lunar Outpost, and the X-33/VentureStar. I supported dozens of working group studies where we formed teams from the NASA centers and would compare our findings. There were the expendable launch vehicle teams and those studying reusable launchers like the Space Shuttle. In summary, what I learned in this assignment was:

- Expendable launch vehicles required a significant and continuous manufacturing operation to fabricate the next launch vehicle. **At no time during my 11 year period of evaluating advanced launch vehicles was any expendable launcher with comparable performance proven to be safer and more affordable that the decommissioned Space Shuttle.**

- The Space Shuttle had become a government "jobs program" and should therefore be privatized to reduce the untenable operations costs.

- The Space Shuttle designers did not plan for flight system upgrades, and as a result upgrades were a major cost during the decommissioned shuttle's lifetime. The next generation shuttle must have replaceable modular sub-systems that can be quickly replaced for rapid launch turnaround or upgrading. Modular sub-systems are a proven cost-saving improvement for commercial and military aircraft.

- Crew escape systems were investigated but were never for a pilotless shuttle launcher. (At this time I was unaware that crew escape pods could have been installed on the Space Shuttle).

- The VentureStar single stage to orbit was the only launch system design that made launch cost reduction a priority. That vehicle was to be a commercially operated venture and was mismanaged (Chapter 4).

- Space Shuttle automated flight systems and commercial launch operations are necessary requirements to reduce launch cost. NASA management believed these requirements to be unsafe for crewed space transportation. They were convinced that only NASA flight controllers and astronauts could be trusted with flight operations of this "national treasure", the Space Shuttle. No serious attempt to reduce flight operations cost of the decommissioned Space Shuttle was ever made.

- There were groups within the aerospace community who believed that the Space Shuttle Program should be canceled because of its

untenable operations costs and campaigned for the return to Apollo-
era expendable launch systems. (The reader is reminded that the
Apollo program was canceled because the cost of launch operations
was unaffordable).

First Letter Crossed the Line

The following is my first memorandum letter to NASA management asking
for an advance launch systems investigation. It wasn't rocket science that
our first priority for advance launch systems should be to reduce operations
cost. General Dynamics, who at the time manufactured the Atlas launch
vehicles, realized that to stay competitive they had to get their launch costs
down. One of their very successful cost-saving endeavors was to develop an
integrated flight design system. I believed their system could be adapted for
shuttle operations and wrote the following:

DF7/89-235   October, 1989
INFORMAL NOTE-
TO: DA2/XXX XXXXXX
FROM: DF75/D. A. Nelson
SUBJECT: Integrated Flight Design System
For some time, I have been convinced that NASA needs an integrated flight
design system. I believe an integrated flight design is required not only
for Shuttle operations, but for all NASA space flight projects. *I did a survey*
at NASA and other aerospace companies and found only one organization that has
incorporated an integrated flight system. General Dynamics (San Diego) has developed an
integrated system to generate flight design products for their Atlas/Titan Centaur project.
Since the system was designed for modular applications, they have been very successful in
adapting it for other projects (spaceplane, advance launch system, space transport vehicle,
etc.). With an integrated system, General Dynamics has reduced their manpower
requirements for flight design products from a parametric analysis system (similar to our
current system) requiring 18,000 hours to their current integrated optimization system
requiring only 7,500 hours. General Dynamics also believes that in 2 years their next
generation program will reduce the manpower requirements to 3,000 hours. In addition

*to manpower reductions, **General Dynamics has reduced the trajectory design time from 4 months to 1 day. With my encouragement, General Dynamics has submitted a proposal (see enclosure) to DM to develop an early assessment integrated system capable of generating the products for Shuttle Advanced Flight Planning (AFP).*** If the integrated system proved to be successful, more detailed application modules could be incorporated later for more sophisticated flight design. DM believes that Shuttle application programs are "too complicated" for a quick look early assessment system, This is a concern which I share; however, I also believe we need to Investigate the potential benefits of an integrated flight design system.* **The flight design system (FDS) developed by MPAD was originally proposed to be an integrated system. I believe it was not successful because of three factors:**

Inadequate hardware

State-of-the-art of available software

Cultural attitude of the flight designers

General Dynamics has proven that the first two problems do not now exist. We need to work on our cultural attitude. The integrated flight system has been proposed and accepted as an RTOP (enclosed). However, funding is not available until 1993. **If NASA is serious about reduction of the cost of flight operations, we should not wait until 1993 to fund this project. Do you have any ideas or comments?**

Don A. Nelson

At the time this was written there were over 800 employees doing the pre-mission Space Shuttle flight design and analyses with an annual cost of $104 million. An integrated flight design system had the possibility of cutting the manpower numbers in half. There were three areas where I had crossed the NASA management "don't go there policy". First, there was no incentive to reduce launch costs, especially reducing contractors, because that has always been a selling point to members of Congress for getting NASA funding in their congressional district. The more people you have on Space Shuttle contracts scattered around the country the more votes you have in Congress. Second, if the program reduced manpower it would cause a morale problem, and third, it would change the way of conducting pre-mission operations and change was a "no no" to management. The Space Shuttle operations management at this time consisted of retired astronauts and former flight controllers, and both groups resisted change and neither have had much if any engineering project development experience. Where engineers live for change, flight controllers and astronauts have the culture of resisting change. If you start changing the way we've done flight operations for decades, someone might even get the idea that NASA didn't need two flight control centers, one in Houston for orbital operations and one at the Kennedy center in Florida for launch operations. However, still today there are two control centers because it creates jobs. Needless to say, the integrated flight design system I proposed was never approved. We had created a massive jobs program, and in 1995 over 21,000 employees were "supporting" the Space Shuttle's missions. NASA's shuttle flight operations were now commonly known as the "Marching Army."

My second try to reduce operations costs in the Space Shuttle program was in early 1990 at a meeting to propose ideas for returning humans to the Moon. I asked to be last on the agenda because I knew what I was going to propose would be very controversial. My proposal was to automate the shuttle and use it as an unmanned freighter to assemble and supply lunar vehicles in low earth orbit. This was not a new concept, as it had been considered in the 1989 *Space Exploration Initiative* except for the automated shuttle. My theory was that automation would reduce mission cost and reduce crew risk since most of the missions would be flown unmanned. It was a backdoor attempt to get management thinking about unmanned shuttle operations. During the presentation I didn't get much flack probably because all of the astronauts had left the meeting earlier. But after my presentation I did get a surprise as one of the senior contractors came up to me and said, "I want to shake your hand…finally someone at NASA has had the courage to say we need to have an automated Space Shuttle flight system."

The Goldin Era 1992 to 2001

Dan Goldin was the longest serving NASA administrator and perhaps the most controversial. He certainly was one of the most technically qualified, having started at NASA as an engineer working on electric propulsion systems for human interplanetary travel. After NASA, he worked in the aerospace industry and became a vice president and general manager for a leading aerospace company. His NASA tenure goal and slogan was to have NASA become, "faster, better, cheaper." The joke around the agency was to pick two, because you can't have all three. In the end, we didn't achieve any of these goals because NASA had become an entrenched government bureaucracy, and any changes were tenaciously opposed.

I believe Goldin knew that getting NASA management to accept change would be his most challenging task. He must have been aware that each NASA center is considered a fiefdom with its own agenda. The biggest 500 pound gorilla in the room was the Space Shuttle Program and its marching army of employees supported by their congressional districts.

Goldin's first effort to get a new management program going was to form Red and Blue review teams to review concepts that supported his approach for a faster, better, cheaper NASA. Employees were encouraged to send their concepts to the Red or Blue team leader. There was no requirement to get your management's approval of your concept before submittal. This implied that any engineers who felt their concepts were not being fairly considered could have it reviewed by an impartial review team. I submitted the automated Space Shuttle concept and it was accepted for further review by the Red team. I was requested to come to the Langley center in Virginia to present in more detail the automated Space Shuttle concept. When I applied for travel funding I was told by Center management that they would not approve the travel. I called the Red team leader and told him I was a no-go for the presentation because I couldn't get the travel authorized. The Red team leader said this is something the new administrator wants because he believes in the review process for new concepts. Later that day I was told the travel had been approved, but not for any overnight stay...make the presentation then get on a plane and come right back to Houston.

The review went well and I was requested by the Red team leader to have a peer review at my center to get feedback on the shuttle automation concept from other engineers. A few days later I got a call from a senior manager on the Johnson Center director's staff. He said he was to be the facilitator for the peer review requested by NASA headquarters. I was to choose six or seven engineers in my peer group and he would choose a like number. In the end we had 15 engineers who agreed to participate in the review. They represented nearly every engineering organization at the center. Following is the letter I sent to the review team:

To: Dist.
From: Don A. Nelson

Subject: Peer Review on a Suggestion for Upgrading the Space Shuttle as an Element in the Next Generation Space Transportation System

NASA's proposed space transportation strategic plan is based on the development of the National Launch System (NLS). This system would consist of a series of expendable launch vehicles developed from a central core booster. Once this system is developed, it would replace the space shuttle system.

For the past three years I have attended numerous presentations and seminars on the development of the NLS. Initially I supported the development of the National Launch System (NLS). However the total development cost for that system to replace the Shuttle capabilities may exceed $25 billion dollars. Studies conducted at MSFC indicate that the life cycle cost of NLS may exceed those of the Shuttle. **Launch costs for the Shuttle now exceed $400 million per flight and there is no disagreement that this system must be replaced.** However, the solution to replacing this system is a very complex puzzle. The space transportation system for the future must not only satisfy our civil needs, but also those of Space Station Freedom and Lunar/Mars.

The Exploration Program Office is proposing the development of an Apollo styled lunar transportation system for a very ambitious development cost of $20 billion. The launch vehicle for this system is predicted to be twice the size of the Saturn V. However, development of this launch system gives no consideration to the infrastructure launch needs of the Agency.

Studies that I have conducted indicate that **an upgraded shuttle system (Shuttle II) and space based transportation vehicles should be considered as candidates for the Nation's next generation space transportation system.** The enclosed documentation defines the design scenario, technology needs, and the development schedule for this system.

The weakness in this concept is the lack of development and life cycle costs. Therefore, I have requested the Administrator to consider funding a study for development of these costs. I believe the funding study will show that when compared to other proposals this concept will offer the most affordable and best operational approach for the Agency.

As a member of the peer review, you have been asked by the Administrator to review this concept and judge its' merits.

My presentation lasted over four hours with questions and debates among the reviewers. It was a give and take engineering review like we had had in the past. My key charts were:

Space Shuttle II Upgrade Considerations

- **Increase fleet size**
  - *Increase spares*
  - *No external changes (same bondlines)*
- **Autonomous Orbiter**
  - *Unmanned flights*
  - *Reduced crew activity*
  - *Reduced crew size (four)*
- **New Ground Facilities**
  - *Operations Processing Facility*
  - *Vertical Assembly Building*
  - *Launch Pad*
  - *Launch Control Center*
  - *Mission Control Complex*

It was the second bullet "Autonomous Orbiter" in this chart that center management was opposed to adopting. Also, I proposed that the Launch and Mission Control Center/Complex should be at the launch site as one facility-which would not be the Johnson Center. That was a "no, no" for senior management, but necessary to reduce launch cost.

Space Shuttle II Upgrade Considerations

Concluded

- *Reduce payloads interfaces*
- *Reduce flight reconfigurations (adaptive avionics)*
- *End-to-end flight design system*
- *Modular systems designed for replacement*
- *Onboard prelaunch and mission vehicle health monitoring*

Note: It is "MANDATORY" that Shuttle II be developed outside of the operational flow of the current shuttle operations.

All of the proposed upgrades on this chart were existing technology and would only require integration development.

Spaced Based "Proof of Concept" Vehicles

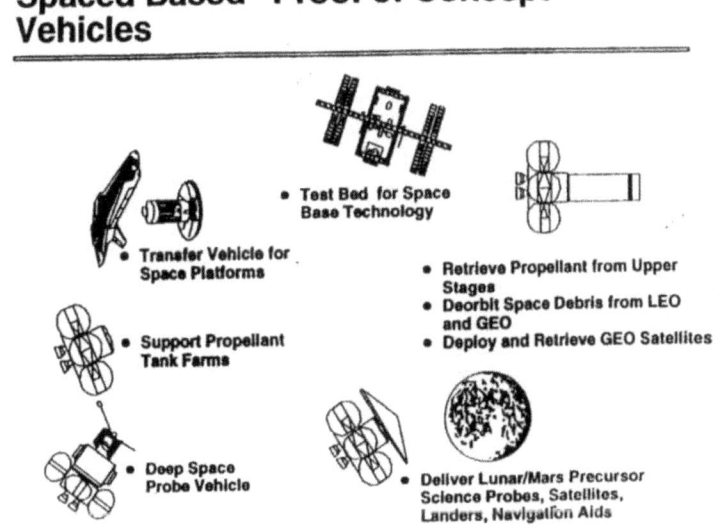

- Test Bed for Space Base Technology
- Transfer Vehicle for Space Platforms
- Retrieve Propellant from Upper Stages
- Deorbit Space Debris from LEO and GEO
- Deploy and Retrieve GEO Satellites
- Support Propellant Tank Farms
- Deep Space Probe Vehicle
- Deliver Lunar/Mars Precursor Science Probes, Satellites, Landers, Navigation Aids

Space Based is another important way for lowering space transportation costs. At the time of this presentation the major technology needed was for long duration storage of propellants. This chart also showed those employees entrenched in shuttle operations that there would be new jobs required for the development of the space based transportation system.

Proposed Development Phases for the STS Elements

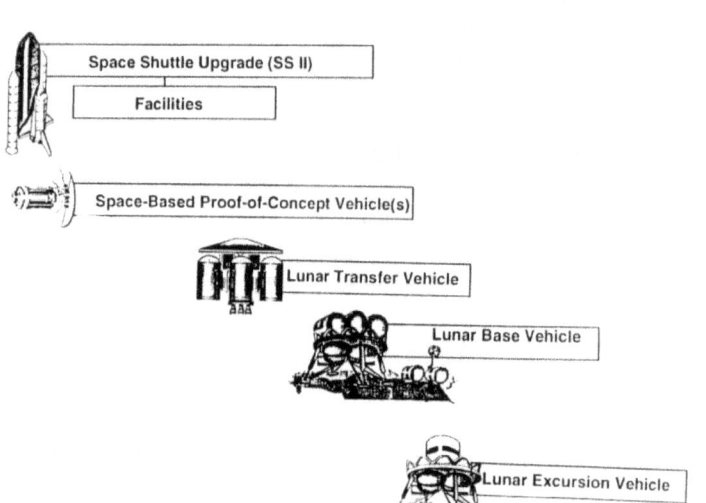

This chart shows that the Space Shuttle II would be the cornerstone for the near earth and deep space systems.

Proposed Mars STS Development Phases

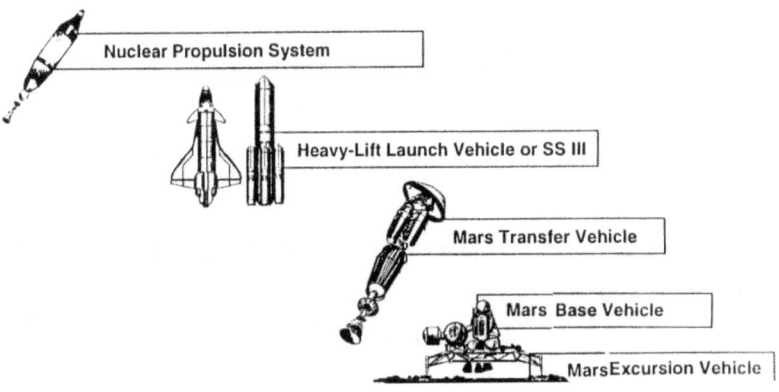

Nuclear Propulsion System

Heavy-Lift Launch Vehicle or SS III

Mars Transfer Vehicle

Mars Base Vehicle

MarsExcursion Vehicle

Needless to say, after the 4 plus hours I was exhausted. I remember one of the reviewers got me a drink of water which was much appreciated. The "senior staff" facilitator stood up and asked the group if they had any more question for me...there were none. He then proceeded to show six or seven charts that in his opinion showed that the automated shuttle concept would never work. He had no data just the standard jargon that the Space Shuttle was a national treasure and the current NASA way to operate the shuttle was the only acceptable arrangement...a government jobs program. Astronauts would never fly on vehicles that didn't have pilots; and, most of all, the nation needed the jobs that the current Space Shuttle operations provided. I couldn't believe what he was doing; the peer reviewers were being told by senior staff that the automated shuttle was not an option for consideration.

The facilitator thanked the reviewers for coming and the review meeting was over. He would later report his conclusions as:

*"It was the **consensus** of the group that many of the authors assessments may in fact be correct but that there was inadequate technical detail to substantiate many of the conclusions. It was also the consensus that the national and NASA management decision process, while incremental in nature and therefore appearing to be disjoint at times, will in due course deal with many of the concerns raised as the basis for the proposal. There was also a consensus that it was difficult to create and implement "grand integrated strategy" in the US system of governance and budgeting."*

The purpose of the review was to get a consensus for requesting Administrator Goldin to *"fund a study for development of these costs"*; also, the technical requirements for automation of the Space Shuttle were well known. As to NASA management "dealing" in due course with the concerns of human space transportation, it is now nearly three decades since this peer review and NASA does not have launch capability to transport our astronauts to the space station except to buy launch service from the Russian government. The reason NASA has never been able to sustain funding for any *"grand integrated strategy"* is NASA has never addressed the issue of launch cost. This is the primary reason one human space program after the other has been canceled…we cannot afford them because of the operational launch costs. The facilitator's "consensus" could be somewhat clouded by the fact that the reviewers were aware that their promotions and future at NASA would be judged by the senior staff that the facilitator represented.

Shortly after the peer review I was called to the Directorate Office and given a "Letter of Caution." These excerpts from the two page long letter stated that:

"Your recent conduct has been unsatisfactory. During the period starting in February 1992, you represented yourself as a "Space-Transportation Systems Advisor" for the Mission Operations Directorate (MOD), you produced and distributed materials which had limited technical verification, and you did not utilize the internal review mechanisms of this Directorate."

"Our judgment, and that of your peers, is that the concepts you have presented are of limited substance and not worthy of further efforts."

"This review gave your concept a fair and impartial assessment in the manner recommended by our new Administrator, Mr. Daniel Goldin."

"The entire Agency is seeking effective direction and options for the future. It is essential that we act responsibly in support of the Agency and its plans. We cannot allow further distractions. The President, the U.S. Congress, and the American public demand professional excellence in our work. I do not believe that your proposals meet this criteria. I personally believe you have abused the privilege of your employment and have compromised our confidence in you and that of your office. I am enclosing a work plan and expect that you will fully comply with its assignments."

If you read between the lines, what this reprimand is really saying is that only NASA can safely operate a manned space flight system, and it can only be funded by a government "jobs program." Anyone suggesting different is not a team player. In any organization except civil service this would have been at letter of termination. However, this was civil service and I had the magic number of 85 points in age and service and could retire at any time with full benefits. I asked the personnel office what the letter of caution meant and was told that if you received three you could be terminated from civil service. Well with two to go I posted the Letter of Caution on bulletin boards around the Center and advised the Red team leader of the risk employees took in participating in Administrator Goldin's peer reviews. I became a lightning rod for getting out information of any employee that felt management was not listening to their concerns, because they knew I would speak out to the highest level of NASA management and even to Congress. That had been the case in my peer review because most of the information in my presentation had come from employees who did not want their name mentioned. My policy still is not to release the names of any of the engineers who give me information.

Administrator Goldin Fumbles

Administrator Goldin's "faster, better, cheaper" quest to get the agency out of its cultural doldrums failed when he elected to not challenge the 500 pound gorilla in the room...Space Shuttle operations. See the following NASA headquarters 1992 release:

Headquarters, Washington, D.C.
September 17, 1992
(Phone: 202/453-xxxx)
RELEASE: 92-154
GOLDIN ANNOUNCES INITIATIVES TO IMPROVE NASA PERFORMANCE
WASHINGTON -- NASA Administrator Daniel S. Goldin today announced a series of broad initiatives and activities to improve the way the agency conducts business and works with its contractors.

"We are committed to strengthening America's belief in NASA as the 'can do' government agency," Goldin said at a speech to the American Institute of Aeronautics and Astronautics (AIAA). "And to remaining an institution that provides inspiration, economic and education benefits to all Americans."

The initiatives represent the work over the past several months of the Red and Blue review teams, which have been **looking at every NASA program on ways to operate faster, better, and cheaper without compromising safety.**

As each of the teams scrubbed through their respective programs, they were tasked to pay particular attention to operational costs that have been growing at an alarming rate.

"We are pleased to report that each team has taken steps to reduce those costs," Goldin said. "We will use those savings to begin planning for new missions, using small spacecraft."

Following are the major initiatives and activities:
NASA'S Program Priorities
*** Shuttle Safety Top Priority -- NASA is working to make the shuttle system safer and more reliable. This includes making investments in new display*

systems to optimize the flight controllers tasks, investing in hardware improvements, improving engine safety, and developing state of the art avionics.

There was no consideration for Space Shuttle automation or privatization to lower the operations cost. Only the "slogan" for shuttle safety was touted. Faster, better, cheaper was dead on arrival and it would be business as usual for the Space Shuttle

First Lunar Outpost Flops

The First Lunar Outpost was an attempt to return humans to the moon and was initiated in 1992. The mission plan was to develop a heavy lift expendable launcher with twice the payload carrying capability of the Apollo Saturn V.

First Lunar Outpost Launcher "Comet"

This was indeed a mammoth launch vehicle. The drawing configuration of the launcher I saw showed that it was too tall to fit into the Vertical Assembly Building. They would have to build the building another twenty of so feet higher or reduce the payload requirement which reduced the lunar stay time.

FLO crew transfers to habitation module

The mission consisted of two launches, where the first launch would be to place an unmanned crew habitation module on the lunar surface. The habitat would support a 45 day stay time on the surface, and the crew would have a rover vehicle for surface transportation. Resupply of consumables (especially oxygen) was one of those issues to be determined.

FLO ascent stage liftoff

The second launch would send a crew of four astronauts and their lunar lander with crew module on a mission to land within walking distance to the habitation module. Although the trajectory profiles and lander were different than the Apollo's, this plan was viewed as a verse two of the canceled Apollo Program. Like the Apollo program, the crew module was limited in its lunar sample payload return capability to 250 kg. This meant that there was no potential for any commercial activity for this launch concept. Without commercial support the launch system was on the same path as Apollo...cancellation.

In November 1991, I attended the First Lunar Outpost presentation to the project director, Dr. Michael Griffin, who was also the NASA Headquarters Senior Manager for Exploration. This was the Johnson Center's introductory meeting for the design concepts of the lunar mission and had an attendance of around 50 engineers, many of which had Apollo experience. It was not a good start, as Dr. Griffin and his deputy manager spent a considerable amount of time discussing the problems the astronauts might have with the 35 foot-high ladder for getting on and off the Lunar Lander and habitation vehicles. The ladder was the least of their problems, but it signaled that their lack of experience in developing launch systems would be the major problem. I was sitting in the back of the room with a group of former Apollo program engineers and overheard their comments on how ridiculous it was to be spending so much time on something so trivial. A few of the back-row engineers were wise enough to know this Super Saturn V launcher concept was dead on arrival. It was at least a $50 billion development program, and each Super Saturn launch could cost from $2 to $5 billion. Although their concept was technically feasible, it was not realistic to believe Congress would ever support this budget-breaking project and they would get little support from the science community. At the end of the meeting Dr. Griffin asked, "Does anyone have any concerns or comments about this concept?" No one replied. Finally, I said, "Would you consider another concept?" Then Dr. Griffin's deputy manager turned around and roared at me, "Where in the hell have you been all day?" He then proceeded to berate me for my lack of understanding of the proposed concept. The meeting ended on that note. There would be no tolerance for opposing viewpoints. However, before Dr. Griffin left the room, I went up to him and said, "You didn't answer my question, would you consider a concept based on inputs from other sources and individuals?" That's one of the advantages of being a government employee who is near retirement. You can say what you believe without worrying too much about the consequences. In the end, this would be one of the reasons for getting another written reprimand from management, but at least Dr. Griffin did agree to review the automated Space Shuttle concept. His review letter and my inserted comments follow:

National Aeronautics and Space Administration

Washington, D.C. 20546

23 January 1992

TO: Don Nelson, NASA-JSC, DA-15

FROM: X/ Senior Manager for Exploration Dr. Michael Griffin

SUBJECT: First Lunar Outpost Plan

Thank you for your comments on the concept for a First Lunar Outpost (FLO) concept which the Exploration Programs Office is currently studying. While your views are almost diametrically opposed to mine, **I think the issues you raise are significant, and therefore worth some attention to a reply.**

I cannot agree that the concept development timeframe for lunar return is too short. The basic modes for going to the moon have been known for thirty years, with at least one of them (lunar orbit rendezvous) known by example to work. Alternate concepts offer both advantages and disadvantages relative to LOR. We need to examine those, put some numbers on the design reference missions, and make some choices. New technology is not needed to go to the moon, and if we don't use new technology, then why should it take a long time to design and develop the next generation of lunar transportation vehicles? (Dr. Griffin failed to understand that the NASA bureaucracy did not exist for the Apollo program.) *There is ample history to document the assertion that you can do almost anything you understand how to do in five years or less.*

I'm not sure I agree that the FLO is much like Apollo, but if it is, is that bad? From my perspective, Apollo was great. Apollo did not "run out of steam"; it was deliberately cut for fiscal reasons, and because, in Caspar Weinberger's words, "it was cuttable". Please see the attached copy of Weinberger's 1971 memo to Nixon. The handwriting on the memo is Nixon's.

I don't know that the science community will, or won't, be satisfied with the transportation system we are designing. Certainly the flight rate is not limited, other than fiscally, to one flight per year.

I share your views on the undesirability of expendable transportation systems. I do not share the view that expendables killed Apollo-. reference the Weinberger memo cited above. But that is a moot point. Certainly, if *** or I knew of any technology that could give us a reusable Earth-surface-to-lunar-surface and return capability, we would use it. Do you know of any? (Dr. Griffin missed point about space based transportation) *If not, then let's get on with what we can do. Nuclear rocketry, once developed, allows reusable LEO-to-lunar-surface and return capability. This will be a nice by-product of technology development for a Mars mission. But it is a couple of decades away.*

Your point about how much Apollo cost is very telling. Clearly, the Nation will not spend that sum today to go to the moon. This is the point we have been arguing, so far unsuccessfully. *We must learn to build space hardware at costs more in line with what we spend on airplane and missile hardware, else there will be no moon base or Mars expeditions.* **It is a matter of our ingrained culture and management style, not of fundamental physics. I'm sorry if I offend, but I believe we (the Government; DOD and NASA) have been the major source of the problem.** (On the culture and management issue Dr. Griffin and I definitely agree!)

I do not understand the point about our "track record" impeding the selling of a lunar transportation system based on current technology. If anything, I would think it works in reverse. I would expect to have trouble selling a scheme based on untried technology.

I'm glad that "many Agency supporters" want us to develop new technology. If so, I wish they'd send money, and send it to Code R, not Code X. Please understand, Code X is in the business of implementing the President's charter, not the business of developing technology other than as it may be essential. As it happens, we expect there will be major technology development efforts in robotics, avionics, medicine, agriculture, etc., necessary to establish a moon base and go to Mars. But our goal is to do it as easily as we can. **** and I are not in business to establish how difficult it is to go to Mars. Improved technology does make things easier -- eventually. But mixing unproven technology with operational missions is exactly the recipe for creating high-cost programs. This is what we're trying to avoid. Possibly you are arguing that we should not return to the moon or go to Mars without undertaking a major technology initiative first. This is one view.

Finally, we are proceeding with only one concept at this stage because we can do only one thing at a time. No matter what path we ultimately select, when we move to implementation we will, ipso facto, have bet all our chips. We hope by that time we will have selected the best approach, but we will select one. From my point of view, a major problem with Moon/Mars is that it has been presented so far only as an array of options, with a high cost associated, and a request to send money. Certainly I cannot guarantee that selecting and advocating a specific, concrete plan will result in the funding to implement it. **But that is the way I'm betting.**

In 1993 Dr. Griffin lost his bet when President Clinton's 1993 budget did not include funding for the First Lunar Outpost program. There was no opposition from Congress or the science community. The cost of the program was never officially announced; however, the production cost for the two expendable launchers, crew habitat vehicle, and crew vehicle was predicted to be $9.6 billion (1992 dollars) plus the operations cost for one mission. That would be over half of NASA's total budget for that year. A sober reminder, "It's the launch cost…" - a fact NASA's human space flight management still refuses to accept.

There was another very serious negative from the mismanaged November 1991 First Lunar Outpost meeting. One of our very promising new engineers told me that he was giving up on a career at NASA. He said what he heard at the meeting convinced him that NASA management was never going to address the major issues needed to get the human space program back on track. NASA has lost too many engineers because of a cultural mindset that has plagued the agency for decades.

Chapter 4   X-33 Mismanaged

Administrator Goldin took another try at the NASA human space transportation dilemma when in 1993 he ordered an in-house study for access to space approaches that could meet the agency's space transportation through the year 2030. The objectives for the "Access to Space" were: to lower launch cost, improve crew safety, and develop a launch system that could compete in the international commercial launch market. By this time the U.S. was losing its dominance in the space launch market to lower-cost foreign launch systems. NASA Headquarters set up three evaluation teams to study the following options:

Option 1 - Identify Space Shuttle hardware changes to improve safety and lowers cost but **make no recommendations for change to the contract and management structure.**

Option 2- Development new expendable launch vehicles.

Option 3 – Develop a new reusable launch vehicle using advanced technology.

Their goal was to define a launch system that would reduce the operations cost by 50 percent, increase flight crew safety, have a reliability that would exceed 98 percent, and reduce the required time for the next launch. With these requirements, Options 1 and 2 were eliminated before the study started. Option 1, the Space Shuttle, would be eliminated because it was the management structure that was the main factor in the excessive launch cost, and the guidelines stated no contract or management change. In other words, don't touch my "jobs program." Also, the Presidential Directive preventing commercial payloads being carried in the Space Shuttle was still in place.

Option 2, for expendable launchers, was also dead on arrival because it required a large manufacturing capability to supply the expendable launch stages, and the labor costs in the U.S. are too high to compete in the international market. There is also a reliability problem in that every launch has the possibility of having an undetected manufacturing flaw which prevents achieving the 98 percent reliability requirement.

Option 3, with a new configuration for a reusable launch vehicle, had possibilities if the advanced technologies could be developed within the NASA budget constraints.

I requested to be on the Space Shuttle Option 1 evaluation team. This was another opportunity to promote the automated Space Shuttle at the Headquarters level and I didn't want to miss that. The week after my request to be on the shuttle team, the Space Shuttle deputy team leader made a point to come to my office and tell me in person that in no way would I be a member of their team. He pointed out that my recent reprimand had marked me as not being a Space Shuttle team player. I wasn't surprised by this and already had a backup plan. After he left the office, I was on the phone to the Option 2 expendable launcher team leader at Marshal Space Center and asked if he needed someone on his team to do the mission operation evaluations. He said he certainly did and would be happy to have me on his team. He needed someone with my background to present their automated mission operation plans. Now my problem was to get travel money from my Center to go to NASA Headquarters to present an automated flight system for another Center. When I submitted my travel order request for the Headquarters' meeting, it was approved with no problems. I think I forgot to tell my management I was on the Marshal Center evaluation team and not the Johnson shuttle team. I did notice that somehow in the Headquarters final summary report my name was listed on the Option 1 shuttle study organization chart for missions operations and not on the Marshal team's list. However, what was important was that I got to plant seeds for automated flight operations. It was what the Headquarters Director wanted to hear and he spent a considerable time asking questions about automated flight operations. As I started my presentation, the Option 1 team Space Shuttle leader and his deputy left the room.

There was no surprise that Option 3 reusable launch vehicle was the choice for proceeding with a contract for the next generation launch system. However the *Access to Space* summary report did reveal a surprise in the Option 1 shuttle evaluation conclusions. They endorsed developing an "uncrewed" (un-manned) orbiter as shown in the following excerpts from their report:

An uncrewed orbiter system was also evaluated. The new avionics system proposed for all of these (Space Shuttle) alternatives would have the increased capability to allow for automation of the ascent and entry functions currently performed by the pilot and commander. The main intent of this new system function was to augment current flights with uncrewed commercial and DOD satellite launches. It was viewed that these missions do not require an on-orbit crew. The Shuttle system could be utilized in this configuration (uncrewed) for general satellite launches. An associated increase in flight rate could result in a "significant reduction in preflight launch cost."

Another is to replace the solid rocket boosters with flyback liquid boosters, which could increase safety and simultaneously improve operations efficiency. The uncrewed orbiter has already had considerable definition, but the flyback booster requires further study to define cost effectiveness.

This in effect confirmed that automated shuttle flight operations would decrease operations cost and *"already had considerable definition"*. The automated shuttle Option was the one NASA Headquarters management should have chosen. It had less development risk, but they would have to privatize the flight operations to conduct "uncrewed commercial satellite launches" and Headquarters' management elected not to fight that battle. This would be another nail in the Space Shuttle's coffin.

Another surprise in the final report was The Option 1 shuttle team had ruled out any changes to their management structure; however, the Option 3 reusable team did study the Space Shuttle management structure and concluded that their automated reusable launch system would reduce the shuttle workforce by over 10,000 people. That's over $1 billion in FY 2017 dollars, but the NASA cultural rule of "don't touch my marching army" prevailed.

The Option 1 shuttle team also reported ***"Providing additional crew escape capability was not recommended due to cost, weight, and center of gravity impacts, and technical risks."*** **However, in another section of the Option 1 shuttle summary report it stated:** *"The uncrewed concept defined by the study resulted in an increase of 10,000 pounds performance and a shift of the center of gravity 26 inches back. This performance gained would have to be balanced with payload* *location or ballast. "* **By removing the orbiter's two pilots, windows and piloting equipment, the "ballast" weight could have been used for crew escape pods.**

The Option 1 shuttle team report confirmed that the Space Shuttle automation could lower launch costs and provide weight margins to permit installing crew modules, but no efforts were made to proceed with these critically needed changes. **Did this failure of due diligence kill the Space Shuttle program and/or worst yet, the Columbia crew?**

Access to Space Reusable Launcher Program

The Option 3 new reusable launcher development program would be management's choice and would be in three phases where the government and the industry contractors would both fund the total project. Phase I would be a fifteen month study to define design concepts for the vehicles. Phase II would focus on one design and build a subscale flight demonstrator vehicle, which was designated the X-33. Phase III was planned to be the development of the operational commercial launch vehicle.

The design guidelines were for a pilotless vehicle which could carry cargo and passengers to orbit without the external tank and solid rocket boosters needed for the Space Shuttle. There were some impressive new technologies which this launch concept would try to incorporate to achieve the single stage to orbit goal, but also some significant manufacturing challenges. The biggest challenge and the impossible one was to get a 45,000 pound payload to low earth orbit on a single stage launcher.

It was a well thought out plan except for two known problems which were ignored. First, sub-scale flight demonstrator aircraft/launch vehicles are notorious for not identifying the manufacturing and flight qualities of the full scaled vehicle. I tried to talk the program manager into building and ground testing a full scale vehicle instead of the sub-scale X-33 flight demonstrator. That decision not to build a full scale ground test demonstrator not only killed the program but nearly resulted in a national tragedy.

Second, a number of my aerospace engineering colleagues and I believed we didn't have the engine to perform a single stage to orbit launch. As it turned out, we were right, and the X-33 sub-scale vehicle never got off the ground, because it was too heavy. Even with today's technology a single stage to orbit launcher is still an impossible goal. Now we were in a situation where the automated shuttle was ignored and the single stage to orbit X-33 was the wrong path.

Competition for Failure

The three major aerospace companies, Rockwell International, McDonnell Douglas, and Lockheed Martin, were chosen to conduct the "proof of concept" Phase I, a fifteen month study. McDonnell Douglas chose the vertical takeoff and lander because they had experience with their DC-X Delta Clipper which had the same configuration. The DC-X Delta Clipper had a number of successful low altitude test flights until a crash ended the project. Reentry from orbit would be extremely complicated because instead of an aerodynamically controlled entry using wings or a lifting body, the entry would use engine propulsion to control attitude and decrease velocity for a vertical touchdown of the vehicle.

Lockheed Martin would use their famous and very successful Skunk Works team to design a lifting body vehicle. Its success depended on the development of composite materials and a new engine configuration called the Aerospike. At the time, composites material had very difficult manufacturing problems and the Aerospike engine never met its design goal. Rockwell International was the third company to participate in the X-33 contest, and their configuration was a winged body very similar to the Space Shuttle, which they built. The Rockwell X-33 used the Space Shuttle Main Engine (SSME), which meant the propellant would have to be carried in the fuselage of the vehicle. Since this was a cooperative government industry project, NASA engineers could participate in the design phase of the X-33 vehicle if the industry partner so desired. I knew many of the Rockwell engineers and had worked with them on the shuttle design and I was asked to be an advisor for the mission operations requirements. Knowing that this was another project that would be canceled, I took the assignment because I was interested in how they planned to design the flight operations systems to reduce launch cost. That process could be applied to the shuttle if the opportunity for automation was ever again considered. Rockwell would reimburse the government for half my salary. This was for the development of a private commercial launch system in which the government was only paying a part of the development cost. This scheme could work today for the design of a commercial reusable launch vehicle. Rockwell's X-33 design immediately had weight problems. There was just no way to get a payload weight similar to that of the shuttle's into orbit in the single stage to orbit configuration. To solve this problem, the Rockwell propulsion guys fudged a little on the performance of the main engine. If they won they could solve the engine performance problem by using the shuttle's two solid rocket boosters. So it would not be single stage to orbit, but, unlike the other X-33 contractors configurations, it would get to orbit...probably without much payload. The Rockwell team really thought they had a shot at winning the Phase II contract. They had in place a manufacturing team with experience in building reusable launch vehicles. Their incentive was if the shuttle was replaced by the X-33 they would have employment as designer and operators of the next generation launch system. They were so confident that they had the winning solution that they were planning to use their retirement fund to finance the Phase III full scale operational vehicles. In the end they lost the Phase II contract to the Lockheed team with the lifting body configuration. Rockwell would eventually be sold to Boeing and most of the engineers would lose their jobs. In my opinion, another big loser was the U.S. taxpayer, who is still paying for unaffordable and unsafe expendable launch vehicles.

VentureStar/Shuttle

Lockheed Martin's Skunk Works team and their lifting body design were chosen to build the subscale X-33 suborbital flight demonstrator and conduct flight tests to confirm the capabilities of single stage to orbit. Their full-scale operational vehicle was to be called the VentureStar. The configuration looked futurist and they had one outstanding public relation team to promote their plan. You could watch on the internet the X-33 being built in the Skunk Works facility and the space cadet community really got excited.

X-33 Range Safety Program…a near disaster.

In the flight program design of any launch system, one of the first priories must be range safety. In other words, if something goes wrong during launch or reentry, where is the debris going? I had worked lots of range safety trajectory issues on other launch vehicles and the Lockheed X-33 really had me concerned. The original flight tests were to launch from the Kennedy Center and fly downrange over the Atlantic Ocean to a landing in Bermuda. The Bermuda government wisely refused to allow the "experimental" X-33 to land on their island. Instead of admitting that there was no place to safely conduct their flight tests, the NASA X-33 project management proceeded to convince everyone that they could safely conduct the flight tests in the continental U.S. My first thought was, "They are out of their cotton picking minds". No one with even the slightest knowledge of range safety flight problems would allow experimental launch vehicle tests over populated areas. All the current launch ranges have safety no-fly lines that if crossed by an errant launcher requires the vehicle to be destroyed before it passes into populated areas. There were no exceptions; if a launcher with a billion dollar payload or shuttle crew crossed over any range safety line into populated areas the launcher would be destroyed. This was not to be the case for the X-33 flight tests. What follows is a series of emails and letters that confirmed the government authorities responsible for protecting the population from unsafe launch tests were incompetent and/or misinformed when they approved the X-33 flight test to be conducted over populated areas. What we had was another case of the responsible not being responsible.

X-33 Flight Test Range

NASA X-33 managers had already signed contracts, and they were not going to let concerns for populated areas stand in their way. They convinced the USAF to let them launch from Edward Air Force Base to Dugway Proving Grounds in Utah and Malmstrom AFB in Great Falls, Montana.

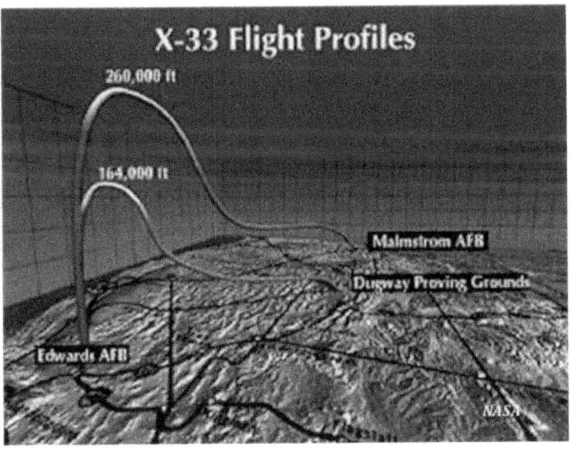

To avoid any public protest of being in a flight test zone, NASA did a public relations tour of the areas where the X-33 would fly over and presented what they called an "environmental impact study." The official statement was:

"To provide for public safety, the range, using Range Safety Programs shall ensure that the launch and flight of launch vehicles present no greater risk to the general public than that imposed by overflight of conventional aircraft."

The X-33 flight demonstrator was an untested rocket and was nowhere near, in any shape or fashion, like a "conventional aircraft." See below:

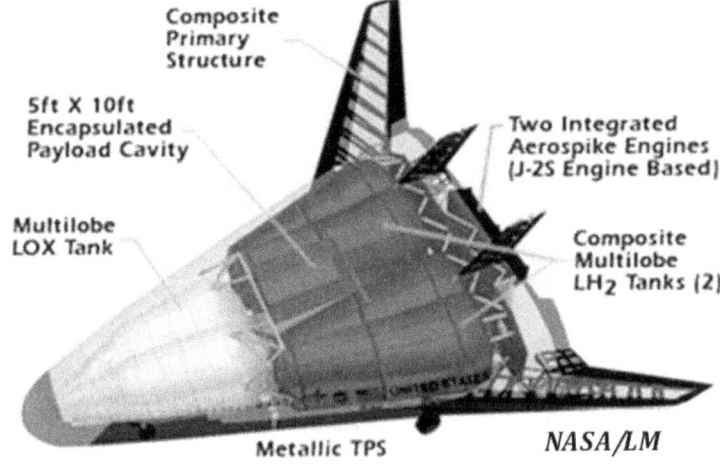

From the configuration of the X-33 it was apparent that the weight concentration was in the back of the vehicle, making it tail heavy and difficult to keep stable in flight. At the time, little was known about the composite structure and the Aerospike engines had never been completely flight tested. The first two flights of the X-33 were to be short distance suborbital flights confined to the boundaries of the Edwards test range. However, it was determined that the aerospike engines would be damaged by the short burn times. So the first test flight would be into populated areas.

The most concerning demon in this vehicle were the two composite liquid hydrogen (LH2) propellant tanks which were known to leak into the inner cavities of the aeroshell body and be trapped. If the hydrogen ignited, the results would become an explosion that produced a deadly rain of scattered debris along the flight path.

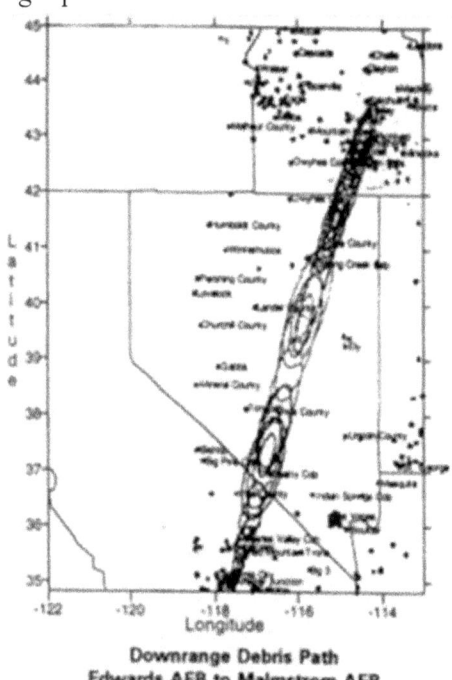

**Downrange Debris Path
Edwards AFB to Malmstrom AFB**

Even with all these obvious safety issues, the X-33 NASA management was claiming that the probability of flight failure was only 1 in 250 flights and therefore safe for flights over populated areas. From all my years in the launch business, I knew this failure analysis was done by someone blowing smoke. It is extremely rare for any new launch system not to have a major failure within the first ten launches. I traced down the organization who conducted the flight failure probability and talked to the "analyst" who came up with the 1 in 250 safety number. **What they had done was base their safety prediction on the launch history of flight-proven launchers and not on a launch vehicle in a test program where the failure rate of 1 in 10 would be expected.** In the 1990 era, the Conestoga, Taurus, Ariane 5, Japan H2, Lockheed Martin THAAD, Boeing Sea Launch Zenit, McDonnell Douglas DC-XA, Boeing Delta III, Brazil INPE, North Korea DPRK, Israeli Shaviyt, and Lockheed Martin Athena have all had one or more catastrophic launch failures in their test phase. In 2015/16 the new highly-praised SpaceX Falcon 9 launcher has exploded twice.

I tracked down the Air Force officer who had approved the X-33 flight test launches from Edwards AFB. I caught him on his last day as he was leaving the service. When I told him that the NASA flight safety number of 1 in 250 was based on operational flight history and not flight test, he said he would have never approved the flight test launches if he had known that was the case. Someone must have passed this information up the chain of command as evident in the following USAF letter:

DEPARTMENT OF THE AIR FORCE

WASHINGTON DC

Office Of The Assistant Secretary

24JUL2000

SAF/MIQ1 660 Air Force Pentagon Washington DC 20330-1660

Mr. Don A. Nelson

CAAE Coordinator

Dear Mr. Nelson

In reply to your letter posing four questions about the X-33 program, the following information is furnished:

The Air Force Flight Test Center commander must approve flight tests from the facility[7] The Air Force will not remove the public from the flight path.

Under DoD D 3200.1 1, the Air Force Flight Test Center commander is the responsible official for public safety in connection with test events conducted at the facility.

Air Force Flight Test Center safety officials are aware that the NASA Environmental Impact Statement was based on operational launch vehicle data and not test vehicle data.

Sincerely

XXXXXX XXXXX

Deputy Assistant Secretary of the Air Force (Environment, Safety, and Occupational Health)

The letter confirms that USAF senior staff was aware that the NASA launch safety number was bogus and they were still unwilling to deny their launch services to the X-33 program! So what happened to no space launch tests over populated areas?

NASA Aerospace Safety and Advisory Panel X-33 Safety Report

I contacted the NASA safety panel to report that the flight safety risk analysis was invalid. When I called their office and reported the error in the flight safety analysis the guy on the phone suddenly starting talking in a whispered tone like he didn't want to be overheard. He asked if he could call back, which he did some time later. He had another person on the line with him who he identified as a NASA safety panel member. I again repeated my concerns about using operation flight history to justify the safety of X-33 flight tests. After the call I thought at last someone of authority would recommend cancelling these flight tests. Was I ever wrong as the following 1998 NASA safety panel report concluded:

Finding #28

The flight profiles for the first X-33 tests will originate in the Air Force Flight Test Center (Edwards) test range and are scheduled to end at Michael Army Airfield on

Dugway Proving Grounds, Utah, which is a storage facility for chemical and biological weapons. Later tests will go from Edwards to Malmstrom Air Force Base in Montana. *While these routes generally traverse unpopulated areas through established military corridors, they also cross several major highways and terminate near vulnerable areas. Also, should there be an unexpected flight termination, the impact could, conceivably, be in a more populated area.* This is particularly true because the destruct mechanism depends on a hard-over flight control signal leading to an aerodynamic breakup that could result in a rather large ground impact footprint. *If this were to happen while unspent propellant is still aboard, the results could be disastrous. Communications failure or command termination failure could exacerbate the situation.*

Overall, the Panel concluded that safety is well served for the present.

Ref: http://history.nasa.gov/asap/1998.pdf

Note: The reader is reminded that the possibility of a flight failure is more likely to be 1 in 10 for the X-33 flight tests and not 1 in 250. Later on in this book the reader will find that NASA continues to report flight failure probability numbers for other human space vehicles that are nothing more than conjecture to avoid cancellation of unsafe crewed vehicles.

The X-33 Flight Demonstrator Fails

In July 1998 *Space News* published this profile of the flight path of the X-33 and the announcement that NASA would post a safety officer to oversee the flight tests.

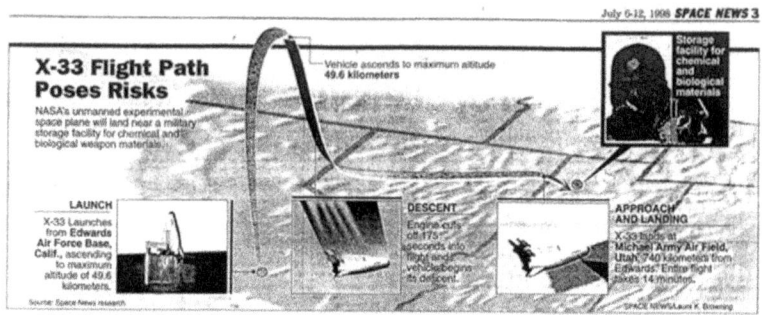

NASA Posts Safety Officer at X-33 Facility

If this "so-called" safety officer had any knowledge of launch vehicle failures, his first task would have been to cancel the X-33 flight tests. NASA's Office of Safety and Mission Assurance was the instigator of this deception to keep the X-33 from being canceled. If this safety officer had done his homework he would have found that the X-33 flight test range included 14 major highways and 16 controlled airways with over 4000 commercial flights in a 24 hour period. What he also should have known was that if guidance command was lost, the X-33 could have gone anywhere since there was no destruct system. A breakup of the vehicle would have sent a predicted 1650 pieces of debris, with some weighing over 650 pounds, into an unknown impact area. If you were this safety officer, would you have canceled the X-33 flight test?

Well, this unknowing safety officer got lucky. On November 4, 1999 one of the hydrogen tanks that was to be flown in the X-33 flight tests broke open in a structural loads tests at the Marshall Center ground test facility, **instead of in-flight**. I believe there were a lot of lucky people that day, because if the hydrogen tank had ruptured in-flight there would have been an explosion. One of the 650 pound pieces of debris could have caused the release of nerve gas at Dugway, and the gas could have drifted to Salt Lake City where fatalities would have been in the thousands.

So why did NASA space flight management put us in this situation? Could it be arrogant ignorance and/or incompetence?

Note: NASA History Project Notes:

"On March 1, 2001, NASA announced that the X-33 Program would not receive Space Launch Initiative funds. The X-33 Program consequently will conclude upon completion of the cooperative agreement between NASA and Lockheed Martin on March 31, 2001."

Chapter 5   Going Public

There was never any question in my mind that the X-33 program would fail. So I decided to introduce the automated shuttle flight operations system to the aerospace community by writing several articles on the concept in aerospace news publications and professional magazines. I felt it would get the aerospace community thinking about the advantages of shuttle automation and also get feedback on any significant technical problems. The first thing I did was get permission from my Center management and the legal office to write the articles. I was informed that it was permissible as long as I stated the article was not an official NASA position and represented my personal conclusions. The reader will note in the following articles this statement is made as was the case for all articles I wrote:

Space News commentary Published January 1994

Don A Nelson

Space Shuttle 2 Answers the Call

Once again a space advisory group — this time the White House Space transportation Group — is addressing the complex issues surrounding U.S. military and civil space launch requirements. The chosen launch system must be affordable, must significantly reduce operational costs, and establish the basis for a long-range strategic plan for a next-generation U.S. launch system.

The advisory group will have less options to consider than previous advisory groups. The National Launch System concept to use expendable launch systems has been discarded because of high development costs of around $15 billion. Proponents of the concept argued that the expendable booster could be used as components for a Super Saturn class vehicle, surpassing the heavy-lift U.S. rockets that carried the Apollo capsules into space. The so-called Super Saturn vehicle proposed for the now defunct first lunar outpost project would have cost nearly $3 billion per flight. What was learned from this defunct project was Saturn-class vehicles are history and the next generation transfer vehicles aimed at reaching the moon and Mars must be space based.

Foreign launch systems can provide an alternative access to space. However, they will be affordable only as long as the United States has an alternative. Current commercial launch systems — such as Pegasus, Taurus, Delta 2, Atlas 2 — can support 5,000 – 20000 pound payload requirements.

Single stage to orbit (SSTO) has recently become a very popular launch concept. The McDonnell Douglas Delta Clipper's successful hops encouraged many of its supporters. NASA's Access to Space studies resulted in a SSTO configuration being recommended to NASA Administrator Daniel Goldin. The SSTO has always been a preferred launch system concept, however it has major technical problems. High-risk technology programs for composite cryogenic booster tanks and tripropellant engines must both be successful to orbit acceptable payload weights.

Also, past studies have indicated that SSTO configurations have serious aerodynamic entry problems. Placement of a cargo bay with sufficient volume is another major design problem. These problems have driven the development costs beyond the $30 billion range.

The space shuttle 2 concept will require significant changes in the NASA philosophy for operating the shuttle. To be an affordable launch system, the space shuttle 2 must have lower launch and operational costs, as well as reduced costs for the external tank and solid rocket booster system. Developing an affordable space shuttle 2 system will require:

• keeping the external configuration of the current shuttle to avoid costly development certifications of the airframe and aerodynamic characteristics;

• developing auto-pilot health-monitoring systems to eliminate the need for pilots and reduce ground personnel;

• creating a hybrid booster system to reduce booster cost by 40 percent, to increase performance, to improve handling safety, and to reduce critical failure;

• ending the lightweight external tank program, freezing the design and eliminating excessive manufacturing costs;

• incorporating modular replacement subsystems that can be changed within hours instead of days;

• developing a long-life, high amperage fuel cell and electrical actuators to reduce hazards and maintenance costs of the auxiliary power system;

• incorporating state of the art thermal protection systems to reduce maintenance costs;

• *establishing a so-called software czar to ensure standard software interfaces; and*

• *installing a non-advocacy review team to monitor, technical progress and development costs.*

The reduction of operations costs is crucial. This will require centralizing both ground and flight operations at Kennedy Space Center. Also, a government corporation should be established to provide launch support, prepare payloads for easy shipment and launch, and use flight crews only for needed pay-load support. Finally, a policy of upgrading launch systems every 10 years would economize by extending the launchers operational status.

A preliminary cost analysis indicates the space shuttle 2 concept could be developed for approximately $9 billion and would reduce the current shuttle operational costs by 60 percent.

In addition to recommending a launch system concept, the space advisory group must endorse a long-range strategic plan for the space transportation system. The plan must use the space shuttle as the delivery system to space based vehicles.

Space shuttle 2 provides avenues to transfer the work force to development projects for space shuttle 2 and space based vehicles. The concept would significantly increase launch rates and provide additional job opportunities in payloads operations.

The space shuttle 2 concept is faster to develop because it uses an existing configuration. It is better because the hybrid boosters significantly improve performance and safety. It is cheaper to develop and operate than alternative launch systems.

Space shuttle 2 is the first step toward providing the United States with an affordable launch system and a direction for the space program. It will not be an easy step. It will take courage, conviction and time to realize that a space shuttle 2 is the only choice.

Don A. Nelson is aerospace technologist at NASA's Johnson Space Center in Houston. The views expressed here do not necessarily reflect those of NASA.

The following article was written for and accepted by a technical aerospace magazine, and expressed the same views as the above Space Shuttle II news article.

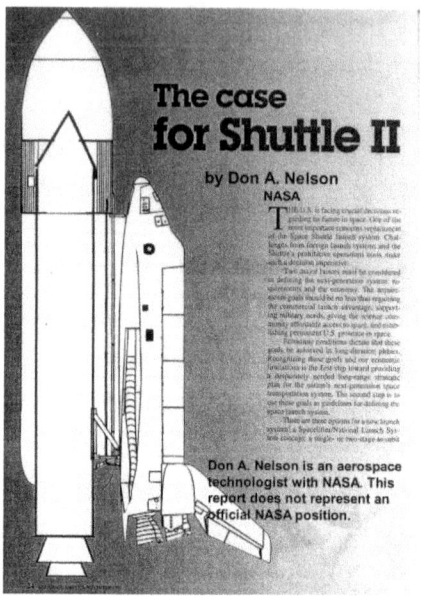

The case for Shuttle II

by Don A. Nelson

NASA

Don A. Nelson is an aerospace technologist with NASA. This report does not represent an official NASA position.

What these articles did was challenge NASA's flight operations procedures requiring their marching army of civil servants and contractors. The articles did get some response in that I got both positive and negative feedback from the aerospace community as the following letters convey:

7 Nov 93

XXX XXXXX

Editor-in-Chief, Aerospace America

370 L'Enfant Promenade SW

Washington. D.C, 20024

Re: The Case for Shuttle II", by Don A. Nelson, Nov. 93 pg. Copy TO: XXX XXXXX

Dear XXXXX:

It pained me to read the subject article in our highly respected and unquestionably authoritative publication. As an Associate Fellow and seventeen year *member of AAIA I felt compelled to respond in the hope that I can repair some of the damage this article will do.*

There are so many egregious errors and self-contradictory statements in the article that I suspect some must be a result of poor editing. However, the total content clearly points toward serious culpability of the author, as well.

I will be specific.

The third paragraph states, without any supporting statement, that "economic conditions dictate that these goals be achieved in long duration phases". With the horrible example of the Space Station Freedom ten year debacle before us, how can anyone advocate more of the same? If no other lesson is learned from "Freedom", let it be that projects with development cycles that exceed the political life of their advocates rarely survive.

The author claims there are "three options for a new launch system" and then cites "single- or two-stage to-orbit vehicles" as one option. Anyone knowledgeable of rocket systems must realize that perhaps the most fundamental argument of this generation of rocket designers is over the issue of single vs. multi-staged rockets.

This technical naivete is continued when he says: "The single or two-stage-to-orbit option has serious problems. Concepts that propose using liquid engine systems cannot meet current Shuttle launch payload capabilities." Here I suspect an editorial error, because the statement does not make sense as written.

Undeterred, he goes on to say: "Thus these systems would have to be nearly twice as efficient as a Shuttle II using the same technology, and would incur the development cost of a complete new launch system. In addition, the launch abort modes for these vehicles have very risky propellant dumping requirements." Here I suspect he meant to deal with single- and multi-stage fully recoverable rockets, but that is never stated. Even then I wonder what his measure of "efficiency" is and how-he arrived at the unsupported assertion that such rockets would have to have nearly twice the efficiency of his proposal? The studies that I have seen and the recent successful multiple flights of the DC-X promise for SSTO's not twice but perhaps 100 times the efficiency of any system based upon the Space Shuttle if measured in cost per lb. to orbit.

On the question of propellant dumping, he apparently presumes that that is a necessary concomitant to a safe abort capability from launch to return of an SSTO vehicle. He is mistaken, because the design that can sustain flight after shutting down an engine can do so from the moment of lift-off without jettisoning anything. That requirement is met by utilizing a number of engines that allows remaining thrust to exceed total weight throughout the flight envelope; a straightforward design concept.

This kind of technical naiveté continues in his faint praise for the NASP, justifying it as having "potential for improved performance" without identifying what the nature of the improvement might be.

His suggestion that removing pilots from a Shuttle II would produce "enormous dollar and manpower savings" is also difficult to accept (even if the prospect of launching other crew members into space without the ability to intervene in the event of a Shuttle anomaly were not bizarre in itself), because all or almost all of those are sunk costs. At best, it would save some personnel costs while enormously increasing the risk in the event of any Shuttle anomaly such as a return to landing site abort.

His trip to technical never-never land continues with the assertion that 'the requirement for manned rated is passe, evidently he believes that "man rating" only applies to piloted vehicles.

Then he adds the profound statement that "All future space vehicles must be designed for maximum reliability". Believe me, Mr. Nelson, if you know how to improve the reliability of any of our current space vehicles you should speak up because their designers surely designed them for the highest reliability that they could.

How to understand the statement under "External Tank" that:" The lightweight tank program should be stopped. Development of the external tank should be stopped and production begun."

This kind of non-sequitur and naivete is continued throughout this paper, inducing in this reader a feeling of acute embarrassment for the author and for my favorite technical publication.

Clearly, this article was not subjected to peer group review, as I cannot picture the distinguished colleagues listed in the masthead Advisory Board allowing such material to appear.

Sincerely,
XXXX XXXXXX

Nelson's Reply:
Mr. XXXXXX sounds like a frustrated DC-X supporter. I too share his frustrations and would like to explore technologies that would lead to a single stage to orbit launch system (SSTO).

However, we must remind ourselves that we are a Nation that is $4 trillion dollars in debt and therefore must spend our very limited research resources wisely. Using current technologies it would take a booster system equivalent to the Titan IV to deliver a Delta Clipper class vehicle with a 15,000 pound payload capability to LEO. To develop a single or two stages to orbit vehicle that could meet the current Shuttle payloads capabilities of 45,000 pounds would require investments in some high risk technology programs and estimated resource funding ranging from $16 to $27 billion dollars. I would urge Mr. XXXXXX and others to take another close look at the more affordable Shuttle II launch concept.

This angry letter writer did make some pointed observations. There are sentences that lack clarity, but that's my fault not the editor's. However, what this letter does indicate is just how much frustration there was and still is in the aerospace community. My frustration was I couldn't get NASA management to conduct an in-depth study of the Space Shuttle II.

The following letter was more encouraging:

XXXXX X. XXXX
Vice President
Business Development
XXXXXXX Inc.
Mr. Don Nelson
D A 15
Johnson Space Center

Dear Don,
Your article in Aerospace America Nov. '93 regarding Shuttle II is excellent. It is specific and, I believe, accurate.
It is refreshing to read a terse hard hitting statement on what I consider to be the single most important issue facing the world space community: **reducing cost-to-orbit.**
I hope people will listen to you.
Keep up the good work.
Sincerely yours,
XXXXX

There were now people in the aerospace community beginning to think of concepts to lower launch cost as the following article confirms:

Aviation Week & Space Technology December 8, 1997
Correspondence
"UPGRADE SPACE SHUTTLE"

Despite its many critics and high operating costs, the space shuttle is probably the best system yet devised for a reusable launch vehicle (RLV). But it was initially miscast in its role as a do-it-all launch vehicle that could be used to provide cheap and easy access to space.

The shuttle is a huge workhorse that is costly to operate and requires a longer turnaround time than originally anticipated. However, it has provided a safe and reliable system for launching manned missions and large heavy payloads into orbit. Rather than discard this highly successful design in favor of a totally different one for an RLV to replace the shuttle, it makes sense to consider redesigning the basic system to improve and simplify its operation to make it more cost-effective. That is the thinking behind the Shuttle 2 concept proposed in 1993 by Don Nelson of NASA's Johnson Space Center.

The idea of a second-generation shuttle fits into the concept of a shuttle family comprised" of a Shuttle 2 for payloads up to 45,000 lb., a downsized kerosene-fueled mini-shuttle for payloads up to 32,000 lb., and successively smaller unmanned versions of the mini-shuttle for payloads from 2,000-16,000 lb. This plan would allow-close tailoring of the size and weight of payloads with vehicle sizes.

Such a family of similar RLVs would be able to launch any payload into orbit at a considerably lower average cost than the present shuttle fleet and an array of expendable launch vehicles operated by the Air Force. In using proven technology, there is the saving of long development costs with the possibility of having to start all over again. *The shuttle family could provide the U.S. with a complete space transportation system for anticipated needs in launch capability through 2030.*
Xxxx X. Xxxxx
Mountain View, Calif.

The following Shuttle II letter from NASA Administrator Daniel Goldin indicates sincerity on his part, but also showed his lack of ability to change the NASA mindset on reducing shuttle operation costs:

National Aeronautics and Space Administration Washington, DC 20546
Office of the Administrator
Mr. Don A. Nelson
Advanced Projects and Planning Office Mission Operations Directorate
Lyndon B. Johnson Space Center National Aeronautics and Space Administration
Houston, TX 77058
Dear Mr. Nelson:
Thank you for your letter transmitting the additional information
on your Space Shuttle II launch concept.
I have forwarded the presentation to the appropriate Headquarters office for review and consideration, *I appreciate the time you took to share your observations with me. It is through letters such as yours that I am able to obtain a variety of perspectives. All views and ideas are integral to our mission, in particularly members of the NASA team.*
Sincerely,
Daniel S. Goldin
Administrator

The NASA Administrator had no place to forward the Shuttle II launch concept for an unbiased consideration and review. It would end up in the Space Shuttle Project Office where it would be filed in their circular file...meaning no action would be taken except to start proceedings for issuing me another reprimand.

Retirement

NASA had set up an excellent program to prepare employees for the challenges of retirement. It was so popular there was a waiting period of six months. Several of my colleagues who had taken the course had advised me to sign up even if I wasn't quite ready to retire. The program suggested many planning options, some of which required several years to put in place. So when I was told my name was on the next retirement class roster I was really looking forward to the class, especially after waiting five months to be accepted. On the first day of the class I had just got seated when I was told I had an important phone call which I had to take. My first thought was there was an emergency at home. It was a relief when I answered and it was from the personnel office. I was told there was a mandatory meeting in the division director's office and to immediately go to the office. I protested that I'd waited five months to get in this class and was told no excuses, mandatory attendance on my part. I asked if I was in trouble again and reluctantly the personnel officer said I was...it was another written reprimand. I got the impression the personnel officer didn't like being a party to what was going on. Since I knew the procedure was to surprise the reprimand recipient so they couldn't be prepared, I went to my office and picked up a little surprise of my own. On arrival in the directorate office I was ushered into their conference room and told to sit at the end of the table facing what looked like a Spanish Inquisition Court. I think they were trying to scare me but I'd been down this road before. Instead of being intimated I was mad...I was mad at my office chief, who was sitting as one of my accusers for not telling me about the meeting. But when I found out what I was being reprimanded for, I really got mad.

When I started to ask a question I was told I couldn't talk, I was to sit and listen as the Director read the reprimand which follows:

National Aeronautics and Space And Administration
Lyndon B. Johnson Space Center
Houston Texas 77058
TO: DF15/Don Nelson
FROM: *****
SUBJECT: Instructions Regarding Correspondence and Proposals Concerning Official Matters

Recently you sent a letter and white paper relating to the next generation launch system to *****, ** the Director of the Johnson Space Center. ***** has forwarded your letter and paper to the undersigned for action. There have also been recent articles dealing with the same general subject and authored by you which appeared in a technical professional journal and in a national newspaper. All of these documents have created the appearance that you are addressing official NASA programs or issues in your official capacity without following NASA policies of obtaining the appropriate technical review or obtaining the full concurrence of your chain of command.

Writings are considered to have been written in an employee's official capacity, if they draw substantially on ideas or official data that are not generally available to the public; or if they deal, in significant part, with any matter to which the employee is assigned or has been assigned during the last year, or any ongoing or announced policy, program, or operation of the agency.

When you prepare written material that states or implies, or it is perceived by NASA and JSC senior management, that your position is taken as part of your official duties or you are using nonpublic information, you must follow the established correspondence and review mechanisms which are intended to assure the quality and veracity of work moving external to the Mission Operations Directorate and the Center. You have been formally and informally cautioned and counseled in the last 2 years about violating these procedures.

For any other type of written material, you should have the appropriate administrative outside employment activity approval. I am unaware of any such approval having been granted to you. The NASA outside activity regulations state that administrative approval should be obtained for writing activities regardless of whether you are being compensated therefor. The appropriate JSC form for processing such a request is a JSC Form 1713.

Under these circumstances you cannot use Government time or equipment, and where you use your official title or position in a scientific or professional journal, you must use a reasonably prominent disclaimer satisfactory to the Agency stating that the views expressed in the article do not necessarily represent the views of the Agency or the United States.

You are again instructed not to send any correspondence that is related to the business of this Agency to anyone within this Agency without the full concurrence of your chain of command. Specifically, any concepts, studies, or recommendations related to *advanced or alternative launch systems will be worked through your chain of command and the Engineering Technology Office before being distributed in any job-related form external to JSC. In addition, **you will not send similar personal and unsolicited correspondence directly to the NASA Administrator**, the JSC Center Director, or any top management of the Agency. **While we are fortunate to work in an Agency that welcomes new ideas,** I must pass on to you the admonition that our leaders expect you, and all of us, to work within and through the NASA organization and that appropriate staff work should be accomplished.*

Finally, if you wish to publish articles that do not involve your official duties, that do not involve nonpublic information, and that do not imply a NASA sanction or endorsement, you should complete the appropriate JSC Form 1713 and have your outside activity approved by the Center.

*I must caution you that this is the final warning on matters of this type. If you violate the instructions of this letter, **it may result in a proposal for disciplinary action.***

XXXXX XXXX

DA/XXXXXXXX:jsc:3/2/94:34522
I acknowledge receipt of this letter.

Don A. Nelson

After I was read the reprimand it was handed to me and was told to sign on the bottom line. First, I asked the personnel officer if I had to sign the reprimand document and they said, no it's not required. Then I pulled out the paper I'd brought from my office, my surprise to their surprise reprimand. It was a reply from the Engineering Director to my request that he review my Shuttle II concept before it was published. His reply was: *"Don, I agree with your thought process. We need a lot more of this kind of thinking...Good Job...Keep it up."* I read the paper tossed it on the table and walked out.

What really made me mad about this reprimand was that I had gone to the legal and directorate offices and asked about publishing the articles, and was told I needed to state in the article this was not an official NASA portion, which I did. I went back to the legal officer and asked why he didn't tell me about JSC Form 1713 and he said he couldn't remember what he had told me. Lesson learned: Get it in writing! If I had known about the required Form 1713, I could have published as "Name withheld at request of author." I went back to my retirement class and nothing else was ever said about the reprimand, but it was my second; and if what I was told by the personnel officer was correct, one more and I would be terminated.

Morale

The morale was really getting bad at the Center, especially if you weren't working shuttle operations. The people in the Engineering Directorate were frustrated because shuttle upgrades were always deferred or if approved were costly to install, and were always behind schedule because no thought had been given to how to upgrade the shuttle in the original design. The Mission Planning and Analyses Division in which I worked the Gemini and Apollo programs was disbanded, and the personnel were distributed to the Engineering and Mission Operations organizations. This meant people lost almost all chances for promotion or salary increases by again being the new guy in the organization. Many were looking forward to retirement and there was a rumor of buyout bonuses. One computer-savvy engineer even wrote a computer program which when the computer was turned on in the morning showed him the month, day, hour, minute, and second that he would be eligible for retirement. He said he had a lot of requests for the program from others wanting to get out of this mess. With no prospect of a new program many were just marking time until they could retire.

Some said that NASA was like a ship without a rudder, with no new human space programs that had a chance of not being canceled. Space Station Freedom was floundering and the International Space Station was still on the horizon, so we reorganized and moved a lot during this time period. What turned out to be the last organization that I would be assigned to was also the worst. The division code was DD and I told people it stood for "doom to die." Management took all the misfits, ill, and trouble makers (like me) and put them in this division. My new job description was still evaluating advance launch vehicle concepts, but I was the only one in this division doing so. To stay in the game I contacted a group at the Marshall Center who were still actively investigating advanced launch systems. We formed a team where I would be the mission operation evaluator and they would be the flight systems evaluators, and once we completed an evaluation we would publish our finding in a government internal note. Most of the launch concepts we investigated were for expendable launch vehicles, and none could meet the capability of the Space Shuttle. However, we did identify new technology that could be incorporated in a Shuttle II launch concept.

One morning I heard one of the engineers in my group screaming out cuss words and went to investigate. He had barged into the division's weekly staff meeting and was waving a paper at the division staff and giving them hell...so to speak. The paper he was waving was an email in which they had sent him an unsatisfactory performance evaluation. In many ways an unsatisfactory performance review was worse than a letter of caution. You're put on probation which could lead to termination. I never had anyone that worked for me that had any potential unsatisfactory performance problems, but the management procedure was to work with the employee to prevent having to write one. No one had said anything to this engineer about his work performance not being satisfactory. So when he got the email I understood why this engineer was so mad. I had never heard of an unsatisfactory performance review being sent by email, but this was the "doom to die" division.

That afternoon as I was getting ready to call it a day and head home, our group leader who had sent the email to my co-worker came into my office. He said he wanted to let me know I'd be getting an unsatisfactory performance report tomorrow. He said that in my work assignment there was a requirement for a report on advanced launch systems which was due tomorrow. Since he believed I would fail to turn in this report he would have to give me an unsatisfactory performance evaluation tomorrow. I had known the group leader for years and had worked with him in Mission Operation Division on shuttle and other advanced systems. I believed he was not happy to be a part of this contrivance. No one had told me about this report. In fact, this was the first time the group leader had ever been in my office to discuss anything about what I was working on. This was a setup to discredit me and get me terminated. The group leader had given me an out because he said the report was due tomorrow. After he left the office I went into action...never give an engineer an out, because I knew the division staff had never read any of my weekly activity reports that I faithfully turned in to the division secretary. One thing about the computer age, you have immediate access to all your data. I took all my activity reports and organized them into report format and in a little over an hour's time I had my advance launch systems report ready. I also had images of the launch vehicles that the Marshal center folks and I had investigated which I included in the report. We had a new color printer in the office, so I made copies of the report with a bright red heading reading, **"ADVANCED LAUNCH SYSTEM REPORT" by Don A Nelson.** I put the report in the chair of the division chief and copies in the inboxes of his staff and went home smiling. Next morning the group leader came into my office with the report in his hand and asked how did I get this done? I just shrugged. Soon after he left, the division chief came in and asked some questions about the material in the report. Guess he wanted to know if I'd written the report. I could have told him anything because he didn't have a clue about advance launch systems.

I had dodged another bullet. It was getting time to think about retirement. My retirement class instructor had said when you retire you must have a plan for what you do in retirement. If you just sit around you will be dead in six months. I had a plan or, as some have said, an agenda.

Sidewalk and Commercial Space Shuttle

In January of 1999 I retired from NASA with two plans for my retirement activity. One plan was to get the city to put in a sidewalk for our road. The road in front of my home was very narrow and dead ended at a very busy street on one end and at the city high school on the other. Kids from the apartments on the busy street had to walk down our narrow road to get to and from the high school. It was just a matter of time until someone got hit by a car. I got on the city planning meeting agenda and presented my case for a sidewalk. I had pictures of kids walking in the street with cars passing closely by them. They listened and replied that the budget was very tight but said they would put the sidewalk on their "to do" list. Someone else was listening…one of the city engineers. In less than a year there was a new sidewalk for the kids. Somehow the engineer found the money that the politicians couldn't.

My other plan was to get this nation an affordable and safe human space transportation system. No more having to tippy toe around the NASA bureaucrats. So far, the politicians and the NASA bureaucrats are still hanging on to their self- interest culture and we've been stuck with more losing human space endeavours, but my hope is that some "engineer" will read this book and somehow find the will and the money to get mankind back on track to the stars.

Chapter 6 Crew Safety
"Crew safety is not NASA management's number one priority!"

I found it very difficult to write this chapter. As I gathered the information and letters that comprise this chapter, I kept thinking if I had phrased a paragraph differently or written earlier, would someone in authority have acted to save at least some or all of the Columbia crew? I assume that this guilt has also plagued some of those who ignored the warning signs of the Columbia's demise. Some will still not concede that if provided crew escape pods this disaster would have been survivable. However, if the reader will stay me they will learn in later chapters that the NASA culture of preaching crew safety as their number one priority still has a hollow ring when faced with the unsafe reality of NASA's current and future human spaceflight vehicles.

Boeing Aerospace Engineer

At an aerospace conference in Los Angeles, I was approached by a Boeing engineer who had worked in the Space Shuttle advance concepts department when it was still a division of the Rockwell Company. He knew of my goal to get the shuttle automated and install crew escape modules. He said if I would go get the analyses of their last automated Space Shuttle study I would find in that study that enough weight from the crew cabin could be removed to permit the installation of crew escape pods. When I got back to Houston I contacted another Boeing engineer who had worked the shuttle automation study and they confirmed that by automating the shuttle enough weight could be removed to permit installing crew escape pods. The report showed that by removing windows, seats, crew cockpit displays and all piloting controls and adding back the weight of the equipment for shuttle automation there would be a positive margin of nearly 2000 pounds. That's enough margin to install four crew escape pods. Now I had information that automation not only reduced mission operation costs it also provided a crew escape system for all phases of flight; launch, on-orbit, and entry. I put together a presentation on shuttle automation with crew escape pods and faxed it to NASA Headquarters. Their email reply follows:

Headquarters Office of Space Flight
X-Priority: 1 (Highest)
Date: Friday June 04 June 1999
To: Don Nelson
From: XXXX XXXXXXXX
Subject: **June 3, 1999 Fax to Office of Space Flight**
Mr. Nelson,

Thanks for your continuing interest in our Space Shuttle Program. I'll answer your fax/letter of June 3 to XXX XXXXXXXXX because Shuttle falls into my area of responsibility.

Safety remains our number one priority, not operations costs. *Human access to space is our primary mission, and upgrades to the Space Shuttle are necessary to support this mission, improve safety, prevent obsolescence, and promote efficiency.*

You offer no practical solutions to the real issues associated with assured crew escape and certainly not in the near term. The reality is that true assured crew escape for all phases of flight will require extensive redesign of the existing vehicle. In short, we don't agree that NASA's commitments for the next two decades of human space flight can be met with your proposed approach.

We don't agree that taking people out of the space program and lowering the Shuttle to the level of expendable vehicles is the right answer, nor is the expense of modifications you suggest justifiable. **The cost-benefit trades have been done and they don't pay back.**

Since you left NASA, great strides have been made in terms of real safety, reliability and cost reduction. We have made, and continue our commitment to making, real changes to the hardware, software and operations - huge differences across the board - and simultaneously the fleet has become much more capable.

Sincerely,

XXX XXXXXXXX
Administrator Office of Space Flight
NASA Headquarters Code M-7
Washington, DC 20546-0001

As the reader will note, the NASA Headquarters manager took one day to evaluate the shuttle automation/crew escape pods presentation. Well maybe not a whole day. Of course they played the safety card; everyone should have known you need pilots on board for safety, especially those who never make pilot errors (ho, ho). Maybe since the NASA Headquarters manager was a former shuttle pilot this might have influenced his thinking. How they concluded that automation and crew pods lowers the reusable Space Shuttle to level of expendables vehicles is beyond comprehension. **I've never been able to get a copy of the** *"cost-benefits trades study* **that doesn't pay back" because there is no evidence they have ever been done.** In my opinion the NASA Headquarters manager was not technically qualified for his position and this decision is one of many based solely on conjecture that drove another nail in the Space Shuttle's coffin, and set back our human space endeavors by decades.

I made another attempt to get NASA management to investigate the automated shuttle with crew escape pods in the following letter to NASA Administrator Dan Goldin:

June 11, 1999
Daniel S. Goldin
NASA Administrator
Subject: Proposed Shuttle Upgrades… Another Management Blunder?

Sir,

I believe that we share a common concern about the possibility that the proposed Space Shuttle upgrades will continue to perpetuate an unsafe and cost ineffective launch system. The Space Shuttle Development conference to be conducted in July appears to be such an attempt.

As a NASA-JSC engineer (recently retired) who worked on the initial design and flight operations of the Shuttle, I am very proud of our accomplishments. However, because of technology limitations we were unsuccessful in our goal to develop a cost effective launch system and the safety provided to the flight crew was and still is questionable. Another Challenger type incident will in all probability have the same disastrous results. With a failure probability of 1 in 423 flights we can expect a Shuttle catastrophic failure on any mission.

Fortunately technology is now available to correct these shortfalls. Unfortunately, Shuttle managers continue to resist these changes. It has been my experience that when these managers are confronted with operational solutions to make the Shuttle safe and competitive they become combative or refuse to listen. Consolidating all operations at KSC under the supervision of a private organization, adopting a "ship and shoot" cargo policy, automation of the flight operation to the level a flight crew is not required, and providing a passenger module in the cargo bay that has abort capability for all phases of flight are mandatory requirements to make the Shuttle launch system safe and cost effective. Unless these mandatory requirements are implemented the Shuttle launch system cannot compete with the launch costs of the EELV's or Ariane V. Failure to adopt these mandatory requirements would also continue to expose the flight crews to a Challenger type incident. It is morally unacceptable to allow this situation to prevail.

A 1997 white paper on a commercial Shuttle "Space Truck" concept based on the above mandatory requirements, recommended that the Shuttle Program Office conduct a comprehensive study on the feasibility of proceeding with the development of a commercial Shuttle launch system. Shuttle managers rejected this recommendation without ever allowing the author an opportunity to present the concept. It is my understanding that a similar concept was proposed by Boeing and rejected. Shuttle managers still insist on continuing the same flight operations philosophy with incremental "evolution" upgrades to the launch system. Incremental upgrades have been proven to be excessively time consuming and costly. The MEDS (glass cockpit), a minor upgrade to remove the mechanical cockpit flight gauges and display their function on a monitor screen has taken eight years, is still not fully implemented, and the cost will substantially exceed a quarter of a billion dollars! Incremental upgrades also introduces serious safety concerns by having to operate a fleet with different vehicle configurations. The commercial Shuttle Space Truck concept with the "mandatory" requirements must be done as a block revision to each vehicle and will eliminate the excessive cost and safety concerns of incremental upgrades.

Unfortunately NASA's XXXXX Administrator of Space Flight does not support a commercial Shuttle Space Truck and has proclaimed that, "We don't agree that taking people out of the space program and lowering the Shuttle to the level of expendable *vehicles is the right answer."(see attachment). It is the right answer if the goal is not to have a cost effective and safer launch system. However, from this statement it appears the goal is to protect what has become a grandiose government jobs' program with a work force that has had a 25% turnover rate in some critical operation areas because of morale problems. The XXXXX Administrator's statement has set the agenda for the July Space Shuttle Development Conference... the space industry will always tell NASA what it wants to hear!*

*Realisticly there are only two options for the next generation heavy lift launch system... a Shuttle Space Truck system with the "mandatory" requirements or the EELV's with an X-38 or X-40 class cargo/passenger upper stage. The single to orbit X-33/Venture Star has been proven to be unfeasible because of weight and propulsion problems. Concepts like the Spaceliner will take decades to develop. Unfortunately there is a third option ... the current Shuttle with incremental upgrades... I believe you said it would cost $80 billion to keep it operating for the next 20 years.... **If this is the route NASA chooses, I predict the Shuttle Program will be terminated after the next Challenger type disaster!***

Your serious consideration of this major issue is requested. I am available to discuss the implementation problems of Shuttle launch cost reduction at your convenience, if you so desire.

Don A. Nelson

Aerospace Consultant

I received no reply to this letter. I believe Administrator Goldin knew he was up against a force that he couldn't control. I've continued to work on the automated shuttle/crew escape pods concept by creating a web page to communicate information to the aerospace community and provide an avenue for contacting me (www.spacetran21.org). What we have is an informal group of aerospace engineers who believe the Space Shuttle reusable launch system provides the best opportunity for 21st century human space travel. I've identified the group as the Concerned American Aerospace Engineers (CAAE). I've taken their inputs and comments over the years and use them to formulate the Commercial Space Shuttle (CSS) concepts. It is my policy to never reveal the source of inputs and contribution to the CSS effort.

Another Challenge to NASA Administrator Dan Goldin

During a question and answer session at an aerospace conference, I asked Administrator Goldin a question on why we didn't have shuttle crew escape modules. He replied it was a weight problem. When I challenged his answer as being incorrect he told me to sit down. My reply was I didn't work for NASA anymore and had the right to challenge him. To his credit he had the Headquarters' Chief Safety Officer contact me later and ask for more information. My reply was as follows:

September 28, 2000

NASA Headquarters

Office of Safety & Mission Assurance

XXXXX,

This reply is the information you requested on my challenge to the Administrator at the AIAA Space 2000 conference. His answer to my question, "Why isn't a crew escape module(s) NASA's number one Shuttle upgrade?", indicates that he has been misinformed. A large payload penalty is not the reason for not installing a crew escape module(s). Even if it was a payload penalty, installing an escape system would still be a mandatory requirement.

The critical issue in installing a crew escape system is that the weight of the system moves the center of gravity of the vehicle (c.g.) to a position where the vehicle becomes aerodynamically unstable. Every 100 pounds added to the flight deck moves the c.g. forward 0.4 inches. This situation can only be solved by removing any flight deck component that is not a mandatory flight requirement. The only significant weight components that are not mandatory are those supporting the piloting function. Automated flight control systems and backup ground monitoring have made the requirement for on-board piloting obsolete. Removing the obsolete non-mandatory piloting components (commander and pilot weight, their seats, forward flight deck display and control systems, escape pole, forward windows, etc.) will provide enough c.g. margin to install four one-person escape modules on the flight deck. When required, additional astronauts can be carried in an escape module placed in the payload bay. This additional weight would be charged to the payload margin.

Automated launch vehicle flight systems are not new technology. Automated ascent flight is a standard procedure for all expendable launch vehicles. The Russian Buran Shuttle test flight was flown without a crew. The X-33 project should provide a state of the art automated flight control and vehicle health monitoring system configuration. The X-38 is being designed for automated flight. The X-40 recently demonstrated a GPS guidance landing to within inches of the runway center line. Automated flight control systems are proven technology.

Shuttle management has steadfastly opposed automating Shuttle flight operations. They have even prevented the testing of the Shuttle auto-land capability. Now they are preventing a crew escape system from being installed by enforcing their opposition to automated flight operations. **The $2.1 billion provided by Congress for Shuttle Safety upgrades is more than sufficient to automate the launch, entry, and landing phases of flight and provide the mandatory crew escape module(s).**

Automated Shuttle flight operations will dramatically reduce the manpower and funding required to support missions. Just deleting the training aircraft and piloting training simulators will be an enormous savings. The reduced operation cost has the potential of making the Shuttle a commercial competitor to the Ariane 5. However, Shuttle automation can be viewed as a threat to the "marching army" of work forces now required to operate the Space Shuttle system. **As I told the Administrator at the conference, "The reason NASA is not installing a crew escape module is political."** *It is certainly not technical. The fate of some future Shuttle crew is now in his hands!*

Don A. Nelson

Retired NASA Aerospace Engineer

I didn't get a reply to this letter. The Chief of the Office of Safety and Mission Assurance was a former shuttle pilot and should have known that the piloting function had to be removed to install crew escape pods. The only result of my challenge was that now all questions at the aerospace conferences are written and screened by the moderator before given to the guest speaker. It's considered by many to be unprofessional to publicly challenge a guest speaker. I can't seem to break myself of this unseemly "unprofessional habit", especially when the lives of a crew are at stake.

January 2001 My First NASA Book

In an attempt to alert the general public about the dire conditions of NASA space human transportation programs, I wrote a book titled *"NASA... New Millennium Problems and Solutions."* It was not what the general public wanted to read about NASA. People like to read good things about NASA, and their astronauts. A space authority professor at George Washington University, whom I asked to review the draft, said in trying to read the book he put it down three or four times before finally giving up. An editor from St. Martin's publishing wrote me a five-page rejection letter with editing recommendations. I decided against taking the time to rewrite the book and self-published the book in 2001. My reason for not taking the time to rewrite was the reports of safety problems with the shuttle were becoming more and more frequent. On page 56 of the book I wrote:

"the question is not; Will we have another Space Shuttle tragedy like Challenger...but When will it happen!"

One More Shuttle Upgrade Study

I cannot recall how many shuttle upgrade studies were conducted, but I can recall the January 2001 study. Crew escape was on the agenda for consideration, but only for show and not for serious study. I asked NASA's lead engineer on the crew escape team why he wouldn't consider flight automation to provide needed weight margin for crew escape pods. His reply was, "I can live with or without crew escape." He might live, but a shuttle crew wouldn't. In talking with another manager he said it would cost a $1 billion to put crew escape pods on the shuttle. His statement was based on conjecture. However, even if correct, the launch cost saving would have paid for the automation and escape pods cost within four years (based on five flight per year and a saving of $200 million per flight). When the crew escape evaluation team refused to consider shuttle flight automation they shut the door on any feasible option to install a crew escape system.

Privatized Space Shuttle

The reader is reminded of the mandated congressional study for shuttle privatization presented in a previous chapter in which the Space Shuttle Program Manager's summary in their September 28, 2001 report was:

*"It is believed that utilization of the Space Shuttle for human access to space will continue through at least 2015 and possibly beyond 2020. The longevity and operational aspects of this program demand a different approach to operational management for the future. **A different management strategy needs to be employed. Privatization of the SSP has the potential to provide significant benefits to the Government.**"*

Once the government is removed from the flight operations, the natural sequence of events would be for the private operator to reduce launch costs that would lead to automation of the flight systems. Crew escape pods would follow and would most likely be a part of the automation flight system change. However, NASA's human flight management would steadfastly resist giving up their operational control of launch operations.

Email Expressing the NASA Culture

In 2002 NASA's human space flight management was once again talking about another "Teacher in Space" program. This would have been an excellent program for promoting science in our school systems if the Space Shuttle had been certified as safe for civilian space transportation. At this time, the flight certification was based on conjecture and not flight proven certification. To issue a warning about another teacher in space program, I had a *Letter to the Editor* published in the Houston Chronicle titled "NASA's Deadly Gamble...Another Teacher PR Stunt." I had again crossed the line when it came to being a team player as the following email from a NASA Space Shuttle engineer's request for "more information" implies:

From: xxxxxxxxxxl@isc.nasa.qov
To: <danelson@xxx.net>
Sent: Wednesday, April 17, 2002 4:17 PM

Subject: more information
I read one of your articles entitled "NASA's Deadly Gamble...Another Teacher PR Stunt" and wondered why you have such suspicion towards an agency which at some point in your 36 year association must have had your loyalty. Maybe you should include in your article information regarding how safe we are compared to other space agencies flying manned missions.

As an engineer working shuttle flight software, the very instrument that would automate flight operations, I would mistrust a system not allowing manual takeover capability. This is a lesson Airbus should have learned decades ago, if you want to talk about saving lives. As far as I know, no manual piloting has ever resulted in damage or disaster. Shuttle upgrades is tasked with making the astronaut more situationally aware. This would contribute to saving not only the crew, but the orbiter, a national treasure, as well.

As someone who has seen the very best NASA ever offered, I find it incredible you wouldn't aid the country in understanding the sometimes insurmountable problems this industry faces...like insufficient technologies which is the reason why no better crew escape module has been designed.

*As someone who works in an inherently unsafe business, my regard for Safety is conscientious and comprehensive. This business risks lives. All of us know it and **I am offended by your implication of sinister politics limiting the quest for enhanced safety.** NASA isn't gambling with lives. Those people choosing to space travel also choose the embedded risks knowing they are entrusting their lives to people like you. What kind of an example are you setting for the new generation?*

XXXXXX XXXXXXX
Orbiter Engineering and Mission Operations Support

My Reply:
From: "Don Nelson" danelson@xxx.net
Date: Wednesday, April 17, 2002 9:26 PM
To: XXXXXX XXXXXX @nasa.jsc.gov
Subject: Re: more information... Good Questions

Your questions are very good! It has been very difficult coming to the conclusion to criticize my former agency. I was there at the best of times. However, yesterday as I listened to the new center director and administrator at the JSC all hands... thinking of what we were and what we've become I felt sadness and frankly I'm ashamed of today's

NASA. As to why, in my last years before retirement, it became painful evident that Shuttle was consuming too much of our funding. Also upgrades to the system never were directed to cost reductions. When engineers tried to input cost reductions they were criticized as not being team players. During this period a study done by Rockwell on the "Access to Space" effort was very encouraging. They proposed an automated Shuttle to be used for commercial payloads. They predicted it would be competitive with the commercial ELV's. The Shuttle would be OV-102 since it was not required for ISS. Shuttle management wouldn't listen.... think about that... That meant that money being used to fly OV-102 could be used for other things... but NASA managers were afraid that the money wouldn't remain in the NASA budget... the Rockwell proposal didn't get anywhere... I tried to get a comprehensive automated Shuttle study conducted and got my hands slapped... but good... It was then I realized that management was not willing to even look at any advanced concepts...

*As a member of MOD Systems Division, I remember the mission control guidance officers feared two things... one an abort and the other was the crew would try to take over and "fly" the vehicle. They trusted the software and not the crew... The sims proved that time after time... Also the next generation manned launch vehicle will be automated... MSFC will see to that... Automated flight was one of the few things that was correct about X-33. As an engineer it is evident that "man-in the-loop onboard" is a weak link for standard flight modes... launch, entry, rendezvous, etc... down mode failure should be ground control... this is the way the Russian do it. Pilots just cannot think fast enough to react. **Think about it... which would you rather have as a passenger on Shuttle... a pilot or an escape pod? You can't have both.***

The New NASA Administrator...Sean O'Keefe

It took President George W. Bush nearly a year to find his NASA administrator. The rumors going around were that anyone qualified knew that the job was a nightmare and they could make three times the pay in the private sector. Finally, in December 2001 Bush chose a career bureaucrat who knew little about NASA and even less about the problems in human space flight. I really don't think it would have made much difference; the Agency was on a one-way trip to disaster with the Space Shuttle management clearing the way. However, I started a letter writing campaign to O'Keefe about the shuttle safety issues. I got one reply from someone in the public relations department thanking me for my interest in NASA. So I tried a different approach by going to the NASA Aerospace Safety Advisory Panel. The NASA safety panel was established by Congress after the deadly Apollo fire in 1967. It was to serve as a watchdog for NASA's management of safety and mission assurance activities.

Letters to the NASA Safety Panel:

January 14, 2002
NASA Headquarters
Aerospace Safety Advisory Panel
Mr. Richard D. Blomberg
Panel Chairman
Subject: Space Shuttle Crew Escape Module(s) Requirement
Sir:

I have requested the NASA Advisory Council to review and recommend to the Administrator, the "NASA Operations Directives for the 21st Century" (see: http://www.xxxxxx.xxx). A key issue in these directives is that without a crew escape module the Space Shuttle is unsafe for human transportation.

This critical issue was brought to the attention of the Aerospace Safety Advisory Panel (ASAP) in September 2000 (see attached memo). To incorporate a crew escape module(s) in the orbiter will require deleting the piloting function. At that time this was a unacceptable political position of the NASA Space Shuttle management. With the confirmation of Mr. O'keefe as the new administrator, this political position may be changed.

Therefore, I am requesting that the ASAP issue a statement on the requirement for a Space Shuttle crew escape module(s). Internal NASA political positions must not be allowed to jeopardize the safety of the flight crew!
Don A. Nelson
Retired NASA Engineer
CC: Code A, Sean O'Keefe

No Reply! Another try:

February 11, 2002

To: *Richard D. Blomberg*
Chairman, NASA Aerospace Safety Advisory Panel
Subject: Space Shuttle Crew Escape Module(s) Study

I am a retired NASA aerospace engineer, who was a member of the flight design and mission control team for the Challenger 51L mission. I will never forget the forty-five minute wait after the Challenger exploded... the time we had to standby to allow the debris to impact before a search and rescue effort could be started. We all knew this would be a hopeless effort, because the crew didn't have an escape system.

We had convinced ourselves that the Shuttle was so safe that the crew didn't need an escape system. We were wrong. However, today that position is still the policy of the Shuttle management... and they are still wrong! Today, we have the technology to incorporate a crew escape module(s), in the Shuttle. However, in my professional opinion the only reason there is not a crew escape system is internal NASA politics... it's certainly not technical!

My efforts to have an "independent" comprehensive study conduct on the escape system have been met with strong opposition because the concept requires deleting the piloting function. Piloting is not a required operation to safely operate the Shuttle... but it is a very emotional issue.

Emotions must not be allowed to put our astronauts and our national space program in harm's way. With the appointment of Mr. Sean O'Keefe as the new NASA administrator comes the opportunity to correct this dire situation. Therefore, I am requesting the Aerospace Safety Advisory Panel to request the NASA Administrator to conduct an "independent" study on the feasibility of a crew escape module(s) for an automated space shuttle flight system.

Every year we wait ... another group of astronauts are exposed to a deadly fate that they shouldn't have to face... Please help prevent another space tragedy!

For supporting information see: http://www.xxxxxx.xxx

Don A. Nelson

Retired NASA Aerospace Engineer

No Reply!

Another Letter to Administrator O'Keefe

The Space Shuttle fleet was again grounded; and, after numerous letters to O'Keefe, I finally got the following letter past the NASA public relations office, but I doubt that O'Keefe ever saw it because the reply was from the Associate Administrator of Space Flight, another former astronaut.

July 11, 2002
NASA Headquarters
Office of the Administrator
Mr. Sean O'Keefe
Subject: Space Shuttle Safety Question
Sir,

At the AIAA Space 2000 conference, I asked NASA Administrator Dan Goldin; "If Space Shuttle safety is NASA's number one priority, why isn't a crew escape module(s) NASA's number one Shuttle upgrade?" His reply that it was too large of a weight penalty, was not correct. He had been misinformed by Shuttle management. It is not a weight or technical problem. It was a NASA political decision not to replace the weight of the piloting functions with the weight of crew escape modules.

At the May 2002 Senate, Technology, and Space Subcommittee, You testified that for Shuttle, "there is no higher requirement than safety." **With the shuttle fleet _again_ being grounded, my question to you is; "If Space Shuttle safety is NASA's number one priority, why isn't a crew escape module(s) NASA's number one Shuttle upgrade?"**

I strongly recommend that you ask the new NASA Chief Engineer, Mr. Theron M. Bradley and Gen. Michael C. Kostelnik to investigate and advise you on this critical issue!

Don A. Nelson
Retired NASA engineer
1407 Moller Rd
Alvin TX 77511

Reply from the NASA Associate Administrator for Space Flight:

National Aeronautics and Space Administration, Headquarters
Washington, DC 20546-0001

August 8, 2002
Reply to Attn of:
Mr. Don A. Nelson
Alvin, TX 77511
Dear Mr. Nelson

Thank you for your letter dated July 11, 2002, regarding the crew escape module for the Space Shuttle. Because the Space Shuttle Program falls within the responsibility of the Office of Space Flight, I have been asked to address your request.

The current Space Shuttle is safe for human operations. *Over the years the Space Shuttle Program has initiated studies for enhancing crew escape options. These studies have always shown that modifying the orbiters to include a crew escape module would provide an increase in the probability of crew survival in the event of a catastrophic event. However, such a design would decrease available crew volume and adversely impact on orbit operational capabilities. Also, the crew's ability to evacuate the vehicle while on the launch pad would be degraded due to the entire crew having to be accommodated on the flight deck.*

In addition, one technology area of significant concern is the design on the crew escape module actuation system. The activation system has to provide sufficient time to escape a catastrophic event, but cannot produce "false positive" indications that would destroy the vehicle unnecessarily. In summary, all previous studies of Space Shuttle crew escape have concluded that the addition of a crew escape module will not lead to an appreciable increase in overall crew safety.

Again, it is due to the your values and commitment to the National Aeronautics and Space Administration (NASA) that makes space transportation not just a dream but also a reality. Your interest in NASA and the Space Shuttle is greatly appreciated.

XXXX X XXXXX
Associate Administrator of Space Flight

My reply comments to this letter dated August 8, 2002:
"The current Space Shuttle is safe for human operations."
My reply: *False, the launch system was currently grounded when this reply was written! The space shuttle is still an experimental launch vehicle; operating with a*

fatal flaw...operating without crew escape modules...even the Russian Buran space shuttle had a crew escape system.

"Such a design would decrease available crew volume and adversely impact onorbit operational capabilities."

My reply: *False, removing the piloting systems provides adequate space on the flight deck for the escape modules. Removing the piloting systems provides the weight margin for the needed four crew escape modules. Four astronauts can conduct any shuttle mission payload requirements. Removing the piloting systems requires a change in NASA's antiquated flight operations philosophies and this is the real issue. Will it require another shuttle disaster to get NASA's flight operations into the 21st century?*

"The crew's ability to evacuate the vehicle while on the launch pad would be degraded due to the entire crew having to be accommodated on the flight deck."

My reply: *False, The entire crew of four can be accommodated on the flight deck and the crew escape modules provide the greatest opportunity of survival. With the crew escape module the crew can be evacuated by crew modules ejection in milliseconds instead of the 15 minutes required to get the tower walkway platform back in position.*

"The activation system has to provide sufficient time to escape a catastrophic event, but cannot produce "false positive" indications that would destroy the vehicle unnecessarily."

Reply: *False positives indications are and will always be a concern. However, state of the art automated flight control systems has significantly reduced the possibility of such an occurrence. The over-riding concern is that a ground flight controller or pilot astronaut will take too long to identify the problem or miss-identify it. Again, will it require another shuttle disaster to get NASA's flight operations into the 21st century?*

One More Challenge to NASA Administrator O'Keefe

Once again in 2002 I would act "unprofessionally" to a NASA administrator at the NASA Alumni League meeting in Washington, D.C. where O'Keefe was scheduled to speak. I was a member of the NASA alumni group and had heard there would be a question and answer session with Administrator O'Keefe after his speech to the alumni group. I wanted his position on the shuttle escape pods issue, so I wrote him some time before the meeting to let him know I would be asking this question and provided some background information so he would be prepared for the question.

At the meeting after O'Keefe's speech in which he talked about shuttle upgrades, the moderator asked if anyone had any question for the Administrator. I raised my hand and the moderator came over and handed me a microphone and I asked why crew escape modules were not a top priority for the shuttle upgrades? His reply was, "I don't know." I thought I would get the standard NASA reply about weight, cost, not safe without pilots, but this guy had never been told anything about crew escape pods. None of my concerns about crew safety and escape pods had ever been called to his attention. So to get his attention I kept the microphone and just stared at him. When the moderator ask for the microphone back I shook my head no and just kept staring. Finally a reporter in the audience stood up and shouted, "give him the damn microphone." Before I handed over the microphone I challenged O'Keefe to find out why crew escape pods were not a shuttle upgrade consideration, then handed the microphone to the moderator and walked out. O'Keefe ignored my challenge. He also ignored the following letter which he probably was never shown:

October 2, 2002
NASA Headquarters
Office of the Administrator
Mr. Sean O Keefe
Sir,
My outburst at the NASA Alumni League meeting *was caused by... you were either were avoiding the space shuttle crew escape modules issue... or you*

actually knew nothing about it! Neither case speaks well for the Office of the Administrator.

A week before the meeting I called your office and asked that you be prepared to answer the crew escape safety question. Also on July 11, 2002, I wrote your office on the issue and received a reply from XXX XXXXX on August 8, 2002. The reply was based on the false premises of the NASA culture... and not an engineering analyses that evaluated removing the piloting function. It was interesting to note that this reply was not the same premise given to Administrator Dan Goldin. At that time it was a weight issue that prevented installing crew escape modules. Now it is a crew accommodation on the flight deck problem. Neither premise considered removing the piloting function.

NASA desperately needs leadership from the Administrator. So once again I ask the question:

At the May 2002 Senate, Technology, and Space Subcommittee, You testified that for Shuttle, "there is no higher requirement than safety." With the shuttle fleet again being grounded, my question to you is; **"If Space Shuttle safety is NASA s number one priority, why isn't a crew escape module(s) NASA s number one Shuttle upgrade?"**

I strongly recommend that you ask NASA Chief Engineer, Mr. Theron M. Bradley and Gen. Michael C. Kostelnik to investigate and advise you on this critical issue!

Don A. Nelson

Retired NASA engineer

I did receive a reply to this letter from the Special Assistant to Administrator who admitted he knew nothing about shuttle operations but insisted that "Crew Safety was NASA's Number One Priority."

The Alumni league president was very upset with me for the rude confrontation of their guest. His emailed:

"We now know the extent of your passion and your fear of a possible repeat of the Challenger incident. However, it does not make your rude confrontation of our guest, Administrator O' Keefe, (namely that he is not concerned about the safety of the Shuttle) any more tolerable.

You have tried to convince NASA Administrators Goldin and O'Keefe in the past that the crew of the shuttle was unnecessarily being endangered but failed to do so. It appears that the engineering and scientific arguments and "solutions" you proposed were not adequate. The fault may well be yours. **The League is apologizing to Mr. O'Keefe for your lack of civility and we trust he is big enough not to hold your performance against us."**

My reply to NASA alumni league president was the following:

"Berating the Administrator, whine in public or be excessively confrontational... I am willing to do all these if I can prevent another Shuttle disaster! Once again... THERE IS NO TECHNICAL REASON THERE ARE NOT CREW ESCAPE MODULES ON SHUTTLE... ONLY POLITICAL!!!

Chapter 7 Columbia Lost

"The Columbia is lost: there are no survivors." President George W. Bush

Request for a Presidential Moratorium

In the opinion of this old NASA engineer, the Space Shuttle at the time was still in the test flight phase of proving its flight worthiness. However, NASA space flight management insisted the shuttle was flight proven and safe. I knew that shuttle flights would never be curtailed for crew escape pods installation until the safety issue was addressed, so I petitioned President Bush to limit the crew size to four, which was sufficient to operate any shuttle mission. The commander and pilot positions would be flown by a crew selection of two to three astronaut piloting teams that would fly all shuttle missions. This way the pilots would be current for piloting and they then could use their training time to prepare for assisting in or performing mission payload specialist activities. Preference for crew selection would be given to astronauts with no young children. I did a survey and there were a sufficient number of shuttle pilots that met this requirement. In no case would anyone be flown as a media public relation event. My request for a presidential moratorium follows:

August 25, 2002
Office of the President of the United States
Mr. George W. Bush
*Subject: Executive Order for a **Moratorium on Space Shuttle Flight***
Mr. President,
I am a recently retired NASA aerospace engineer and it is my duty to inform you that our space shuttle astronauts are in eminent danger. Your intervention is required to prevent another catastrophic space shuttle accident. NASA management and the Aerospace Safety Advisory Panel have failed to respond to the growing warning signs of another shuttle accident. Since 1999 the launch system has experienced the following potential disastrous occurrences:
July 1999 - Space Shuttle Columbia delayed by hydrogen leak.

December 1999 - Space Shuttle Discovery was grounded with damaged wiring, contaminated engine, dented fuel line, and paper work errors.

January 2000 - Space Shuttle Endeavor is delayed because of wiring and computer failures.

March 2000 - Space Shuttle Atlantis main engine must be replaced because of paperwork errors.

August 2000 - Inspection of Space Shuttle Columbia reveals 3,500 defects in wiring. Wiring defects plague entire fleet.

October 2000 - The 100th flight of the space shuttle was delayed because of a misplaced safety pin and concerns with the external tank.

April 2001 - NASA failed to keep adequate watch on safety operations of a major contractor.

July 2002 - The inspector general reports that space shuttle safety program not properly managed.

April 2002 - Hydrogen leak forces scrub of the Atlantis flight.

August 2002 - Shuttle launch system grounded after fuel line cracks are discovered in all the fleet!

Mr. President, as you are painfully aware NASA management has been lacking for a number of years. Unfortunately, your new NASA Administrator has failed to recognize the eminent space shuttle danger and has accepted the consul of the pre-existing NASA shuttle management. These managers still pursue a management philosophy that has stagnated the safety upgrades efforts and perpetuates the staggering launch costs.

The space shuttle or any space transportation vehicle without crew escape modules will never be safe to transport humans. To incorporate crew escape modules in the space shuttle requires that the piloting function be removed from the vehicle. Unfortunately, the background of the shuttle management is that of former flight controllers and astronauts. They have been trained to never trust automated flight control systems. Therefore, they are adamantly opposed to automation of the space shuttle. Efforts by NASA engineers and contractors to automate the shuttles are met with stern rebukes and reprimands in some cases.

Mr. President, to prevent another shuttle disaster it is requested that an Executive Order be issued that places a moratorium on space shuttle operations. This moratorium must limit shuttle missions to flight crews that do not exceed four members. The moratorium must remain in effect until crew escape modules can be incorporated.

This moratorium will serve as a catalyst to kick-start the resisting NASA management into action. The lives of our astronauts and the future of our space program must not be ignored. The warning from the Thiokol engineer was ignored and the Challenger exploded. The terrorist training warning from FBI agents was ignored and we had the 9-11 disaster. **When the next shuttle explodes...and Murphy's Law says it will, we can exclaim with pride a loud "YES!" as the crew escape module carries our astronauts to safety... or if this moratorium is ignored...we can watch in horror and shame as the astronauts face certain death**

Don A. Nelson

Retired NASA Aerospace Engineer

I had sent my letter to President Bush by way of his Office of Science and Technology Policy because if I sent it to the White House it would be lost in the thousands of mail received every day. As it turned out, the letter never got to the President's desk, as the following reply verifies:

EXECUTIVE OFFICE OF THE PRESIDENT
OFFICE OF SCIENCE AND TECHNOLOGY POLICY
WASHINGTON, D.C. 20502

December 4, 2002
Mr. Don A. Nelson
1407 Moller Road
Alvin, Texas 77511

Dear Mr. Nelson:

Thank you for bringing to my attention your concerns regarding the safety of the Space Shuttle and its crew. Following receipt of your letter, **my office met with NASA** *officials to discuss your concerns and the Space Shuttle safety program. NASA places a high priority on safety and has instituted a program of developing and implementing*

*safety upgrades to reduce the risk to Space Shuttle crews. Based on these discussions. **I do not think that it is appropriate for the President to issue a moratorium on Space Shuttle launches at this time.***
Sincerely, *XXX XXXXX Director*

The Office of Science and Technology Policy is the president's oversight for the nation's science and technology activities. It should have been the responsibility of the office to inform the President of any event in the science and technology arena that has potential harm to the nation and its citizens. When their only effort was to ask "NASA officials" if there was a shuttle safety problem, they failed their President and the Columbia crew.

My reply to their letter of December 4, 2002:

December 21, 2002
Executive Office Of the President
Office of Science and Technology Policy
XXX XXXXX, Director
In reply: To your letter of December 4, 2002
Mr. XXXXX:
I assume that you are aware that another propellant leak has occurred on the space shuttle!
I assume that you are aware that there has never been a launch vehicle that has not had multiple catastrophic failures.
I assume that you are aware that the space shuttle is the only launch vehicle carrying humans to be operated without a crew escape module.
I assume that you are aware that there are no technical or operational problems that would prevent automating the space shuttle flight operations and installing crew escape modules.
I assume that you are aware there has never been an independent unbiased evaluation for the justification of crew escape modules.
I assume that you are aware that NASA's human space flight management's refusal to automate space shuttle and International Space Station flight operations will result in

the minimum total cost of that project to be $117 billion by 2012. (See: http://www.xxxxxx.xxx/#Petition)

I assume that you have informed the President that the request for a moratorium has been denied and his administration is accepting the responsibility for the fate of the space shuttle crews.

It is urgent that your office immediately initiate an independent unbiased evaluation for space shuttle crew escape modules!

Don A. Nelson   Retired NASA Aerospace Engineer

No Reply!

February 1, 2003...the morning I cried.

On this fateful morning I was at an auto service facility getting the state inspection for my truck when my wife called to tell me they had lost contact with the shuttle, Columbia. About the same time an announcement came over the television in the waiting area of the automotive shop. It was like someone had hit me hard in the stomach. As I drove home I listened on the truck radio to what information they had at the time. All communications with the Columbia had been lost. Debris believed to be from the shuttle had been reported to have been found in East Texas. There was no doubt we had lost the Columbia and its crew. I pulled the truck into my driveway and just sat here... and then I cried.

I still tear up today when I think how I...how we... failed this crew. For those who say human space flight is a risky business and the crew knew that and were willing to accept the risk, I say they are correct, but did we have the right to put them in harm's way, when there were steps we could have taken to significantly reduce their risk? We failed this crew; I failed to convince NASA's human space flight management that crew escape pods could be installed on the shuttles and NASA management refused to listen. In this year of 2017 I am still failing and NASA management is still putting astronauts in harm's way.

President Bush was never was made aware of my warning letter to him. His Director of Science and Technology Policy trusted NASA's human space flight management in that *"NASA places a high priority on safety."* Did that misplaced trust cost these astronauts their lives?

The above photo of the Columbia crew was reported to have been recovered from the Columbia debris. I was never able to verify that it was, but the Columbia accident report concluded:

"it is irrefutable, as conclusively demonstrated by items that were recovered in pristine condition whose locations were within close proximity to some crew members, that it was possible to attenuate the potentially hostile environment that was present during CM (crew module) break-up to the point where physically and thermally induced harmful effects were virtually eliminated. This physical evidence makes a "compelling argument" that crew survival under environmental circumstances seen in this mishap could be possible given the appropriate level of physiological and environmental protection."

Even with their own compelling argument, the Columbia Accident Investigation Board refused to recommend that crew escape pods be investigated. The crew died from blunt force trauma and hypoxia…in other words they were beaten to death and suffocated. Would crew escape pods have saved their lives?

Media Blitz

When the media discovered that I was the retired NASA engineer that had warned President Bush about a shuttle disaster I was bombarded with requests for interviews from newspaper, radio, and television.

—— *Original Message* ——

From: xxxxxx@martyqriffin.com

To: <danelson@wt.net>

Sent: Monday, February 03, 2003 2:38 PM

Subject: **Interview request**

Hi Don

I heard you on Dateline and thought the information that you were **providing was incredible and needed to be heard by more of the public.** *I am the executive producer of a consumer help show, The Marty Griffin Show, on KRLD (CBS radio) in the DFW area. I was hoping you would be willing to join us tonight about 8:20pm central time via the phone to discuss some of the safety concerns that you have had about the program. Please let me know if you are available as soon as possible either via this email or you can call me at 214.826.xxxx. If you are interested we would call you to confirm and then call you this evening shortly before your interview slot So we would need your evening number. I suspect we would have you on for approximately 20 minutes or so. Again please let me know if you are available either way.*

I did interviews with Fox News, Larry King Live, Tokyo Broadcast System, Los Angeles Times, Inside Edition, ABC, US News & World Report, Associated Press, CNN, America Live talk show, The Today Show, Washington Post, CBS News today, Canada TV, Bill O'Reilly, CBS News, Good Morning America, and many more. The end result was all the interviews were just sound bites to fill a producer's program time and I convinced no one to take any action to correct the problems in NASA's human space flight program. I had a high school teacher who said only ten percent of the adult population in this country is willing to try to change the way our society works if there is a problem. I disagree with that estimate; I think it less than 1 percent.

I finally stopped doing interviews after years of requests. My repeated challenges to media producers to do an investigative report on the management problems in NASA's human space programs went ignored. Our human space flight programs continue to flounder and we still have depressing crew safety issues. It is my hope that this book will awaken them to the dire consequences of their continued silence.

The Columbia Disaster Hearing

I'll let the following Congressional hearing report inform the reader on the NASA Administrator's position for the shuttle crew escape system. The reader is asked to recall my staring challenge to Administrator O'Keefe back in September 2002 at the NASA Headquarters alumni meeting, when I asked the same question as Sen. Boxer.

Columbia Disaster Congressional Hearings:
Feb. 12, 2003

Sen. Barbara Boxer: **So I want to know how you feel about this array of facts. First of all, do you agree that the time is past due for the implementation of a more capable crew escape system?** *And if you do, why haven't we seen more done about it?*

NASA Administrator O'Keefe: my understanding is that the analysis that went on a couple, three years ago, following that particular set of reports, of the options all led to a series of technical modifications to the shuttle which would have increased its weight dramatically--its operations, its maneuverability--and so therefore were deemed to be marginal improvements in safety that could be attained--if at all--and yet dramatically increased weight, which would have compromised the safety upon orbit capabilities.

Cong, Ralph Hall: Mr. O'Keefe, you heard my opening statement. And I'm quite frankly disappointed that 17 years after the Challenger accident, so little attention has been given to developing crew escape systems for our astronauts, whether they're flying on the shuttle or whether they're in the Space Station.

O'Keefe: But with regard to the specific crew escape efforts, recall that since Challenger, there have been a number of operational changes made. There is an egress system that was put into place right the Challenger accident that was part of the Rogers Commission recommendations--that ultimately stem from it, I should say--that we put into place, that now still exist to this day.

Once launched though, there is a number of different approaches that have been proposed, examined, reviewed and all of which added significant amounts of weight, I am advised, to the overall effort and so, as a consequences, were viewed to be technically infeasible.

So we take every precaution in this process, in order to assure that, all the way through ascent, that every possible opportunity is there, as much as possible. But again, the idea of an escape system was looked at, examined very thoroughly. And the conclusion was that the weight factor would almost be prohibitive, in terms of its technical clarification.

So we will continue to look at that. We will go back and look at it again, you bet. In light of this circumstance, we really do need to focus entirely on what all the alternatives are.

And I guarantee you, sir, we'll make that part of our effort underway now as part of this November amendment that is before the Congress, to consider for the 2003 program, that we'll factor that into the equation and proceed as appropriate.

NASA Administrator O'Keefe failed to tell the hearing members about crew escape pods and that there was sufficient weight margin to install them if NASA would automate the shuttle flight systems. Perhaps he still didn't know, but if crew safety was NASA's number one priority he should have at least been concerned when he was confronted by this aerospace engineer in the September 2002 meeting. **Congress would later fund $15 million for a shuttle crew escape study, but none of that money would be used to investigate the crew escape pods.**

Columbia Accident Investigation Board (CAIB)

The CAIB would put another nail in the shuttle's coffin which would eventually send them to museums. In my opinion, this investigative team lacked the technical qualifications to conduct the Columbia disaster investigation with the exception of one member, Gen. Duane Deal. First, they never understood this was not an accident; it was a preventable disaster. Second, they refused to recommend a crew escape pods investigation which was the only feasible solution for improving crew safety. Again, in my opinion the majority of this board had already concluded that the Space Shuttle was unsafe for human space transportation. However, I sent a presentation to the investigation board that defined the need for crew escape pods. Their reply follows:

CHAIRMAN
COLUMBIA ACCIDENT INVESTIGATION BOARD
16850 SATURN LANE, HOUSTON, TX
Ser CAIB/009 5
5 Mar 03
Dear Mr. Nelson
Thank you for your letter of 3 March.
I appreciate the report you provided. Your letter has been provided to our independent technical group, and they will be in touch with you if further information is needed.
Once again thank you for your input.
Sincerely,
H. W. GERMAN
Admiral, U.S. Navy (Retired)
Chairman, Columbia Accident Investigation Board

This was a "don't call us, we'll call you" brush off. I knew some of the members of the investigation board and they knew my position, or as the following letter from the NASA Alumni League states, they knew I had an "agenda":

Minutes of NAL/JSC Chapter Board Meeting

April 11, 2003

Space Center Houston

Norm has been contacted to assist in arranging a Columbia Accident Investigation Board Meeting with NAL representatives. Meeting will be for Shuttle Historical Perspective inputs. Three or four NAL members will first establish who they are and what they represent, and then the Board will ask questions...Aaron Cohn has accepted Norm's invitation, and he is waiting for responses from Owen Morris, and Bob Thompson. Other suggested names were Max Faget (health question), Milt Silveira, Sy Rubinstein, Dale Myers, and George Jeffs. **Don Nelson asked that he be included on the panel. Don was informed that he had an agenda and his name would not be submitted.** *That his inclusion would not be responsive to the CAIB request. This action was not meant inhibit Don from presenting his thoughts to the CAIB in another forum. Meeting is at the Clear Lake Hilton, April 23, 9:00am to 12:00 noon.*

The results of the investigation board can best be summed up by the following excerpts from their final report:

"The attitudes and decision-making of Shuttle Program managers and engineers during the events leading up to this accident were clearly overconfident and **often bureaucratic in nature."**

"The changes we recommend will be difficult to accomplish - and **will be internally resisted."**

"NASA's current organization does not provide effective checks and balances, **does not have an independent safety program***, and has not demonstrated the characteristics of a learning organization."*

"If NASA will accept this prescription and take the "medicine" prescribed, we may be optimistic regarding the program's future; if, however, NASA settles back into its previous mindset of saying, 'Thanks for your contribution to human space flight,' summarily ignoring what it chooses to ignore, **the outlook is bleak for the future of the program."**

It is the following two statements that lead to the demise of Space Shuttle Program:

"The Space Shuttle "has never met any of its original requirements for reliability, cost, ease of turnaround, maintainability, or, regrettably, safety."

"Based on NASA's history of ignoring external recommendations, or making improvements that atrophy with time, the Board has no confidence that the Space Shuttle can be safely operated for more than a few years based solely on renewed post-accident vigilance."

The investigation board made no recommendation as to what launch system would meet the requirements for "reliability, cost, ease of turnaround, maintainability, and safety." The reason they didn't was because there were none better than the Space Shuttle. The results have been the decommissioning of the Space Shuttle and billions of dollars of wasted money on obsolete, unaffordable, and unsafe expendable launch systems which don't meet the requirements for "reliability, cost, ease of turnaround, maintainability, and safety."

If the following section on crew survivability from the accident report makes the reader mad as hell, that was my intention:

Columbia Accident Report (Appendix G.12 Crew Survivability)

CONCLUSIONS:

Acceleration levels seen by the crew module prior to its catastrophic failure were not lethal. LOS occurred at 8:59:32. **The death of the crewmembers was due to blunt trauma and hypoxia.** *The exact time of death - sometime after 9:00:19 a.m. Eastern Standard Time - cannot be determined because of the lack of direct physical or recorded evidence. Failure of crew module was precipitated by thermal degradation of structural properties that resulted in a catastrophic sequential structural failure that happened very rapidly as opposed to a catastrophic instantaneous 'explosive' failure. Crew module separation from the forward fuselage is not an anomalous condition in the case of a vehicle loss of control as has been the case in both 51-L (Challenger) and STS-107 (Columbia).*

SUMMARY

It is irrefutable, as conclusively demonstrated by items that were recovered in pristine condition whose locations were within close proximity to some crewmembers, that it was possible to attenuate the potentially hostile environment that was present during CM break-up to the point where physically and thermally induced harmful effects were virtually eliminated. **This physical evidence makes a compelling argument that crew survival under environmental circumstances seen in this mishap could be possible given the appropriate level of physiological and environmental protection.**

In other words, the crew escape pods could have saved their lives. The Columbia accident board ignored their own findings that the debris found in "pristine condition" from the crew cabin made a "compelling argument" for crew escape pods. Instead of recommending that crew escape pods be installed on the shuttle they reported: *"the Board has no confidence that the Space Shuttle can be safely operated for more than a few years."*

A Return to Flight Task Group was formed to insure that NASA implemented the recommendation of the CAIB. Since the board refused to recommend that crew escape pods be installed, the task group refused to consider the escape pods. See following letter:

RTG TG 04-040   June 4, 2004
Mr. Don A. Nelson
1407 Moller Road
Alvin, TX 77511

Dear Mr. Nelson:
Thank you for making us aware of your concerns. The Return to Flight Task Group is committed to doing its part to help ensure the safe return of the Space Shuttle to flight.
The Task Group is chartered to perform an independent assessment of NASA's actions to implement the recommendations of the Columbia Accident Investigation Board (CAIB) as they relate specifically to the next Space Shuttle flight. Our job is to determine whether NASA's plans and actions meet the intent of the CAIB return to flight recommendations, but not to suggest specific remedies.

Some of the issues you raise in your letter - specifically those related to the adequacy of the CAIB recommendations, automating the Shuttle, and **consideration of a crew escape system -are beyond the scope of our effort. Per its charter, the Task Group will not attempt to assess the adequacy of the CAIB recommendations.** *However, your concerns about detrimental flight schedule pressure were shared by the CAIB and addressed in its return to flight Recommendation 6.2-1. I want to assure you that the Task Group has and will continue to carefully study, publicly deliberate, and report on NASA's progress in complying with this and all CAIB return to flight recommendations. Thank you again for your input.*

Cordially,

XXXX XXXXXX

I wrote a letter back to the task group recommending they resign in protest over being stopped from performing an independent assessment of the crew escape pods. They ignored the letter. The Return to Flight Task Group blindly followed their orders and in doing so failed to protect a future crew. To my surprise there was one group that did have the courage to challenge NASA space flight management and resigned.

NASA Safety Panel Resigns

All nine members of the Aerospace Safety and Advisory Panel resigned, reported to be because of sharp criticism from the Columbia Accident Investigation Board and Congress. However, the following excerpts in a *New York Times* article in September 2003 give a different viewpoint on the reason for the resignations, at least for some members:

The panel chairwoman, Shirley C. McCarty, said in a telephone interview that members had "a very big sense of frustration."

Ms. McCarty, an aerospace consultant, said her group had become even less effective after its annual meeting, shortly after the Columbia accident, **when it advocated installing a crew escape mechanism on the shuttle and reorganizing the space agency so the safety officer for a flight would be more independent, and would not report to officials responsible for flight operations.**

NASA was not receptive, she said, and "this set a stage for less effectiveness."
Ms. McCarty said shuttle crews would face risks similar to those of the Columbia unless an escape system was developed.

Note: The reader will recall I had requested the NASA safety panel consider the crew escape pod in 2002 and received no reply.

PRESS RELEASE
Date Released: Tuesday, September 23, 2003
Rep. Hall
Rep. Ralph Hall Statement on ASAP Resignations
Congressman Ralph Hall (D-TX), Ranking Democratic Member of the House Science Committee, today released the following statement:
"The mass resignation of the members of the Aerospace Safety Advisory Panel (ASAP) sends a strong signal that, despite the useful and important service that they have provided over the years, their advice has rarely been heeded. Simply changing its membership will not improve ASAP unless NASA and the Congress are willing to dedicate the resources and effort necessary to implement the Panel's safety recommendations."

"Earlier this year, ASAP members strongly urged NASA to install a crew escape system on the Shuttle. Unfortunately, NASA has not embraced this advice, instead focusing on a new crew transfer vehicle that may not be available for many years. NASA should not ignore ASAP yet again. It should commit itself to carrying forward the Panel's very important recommendations on Shuttle crew escape and safety."

NASA Office of the Inspector General

In December 2003, I contacted the inspector general's office and requested they investigate NASA negligence in failure to consider crew escape pods on the shuttle. I was told by a staff member that they were "too busy" for that type of investigation. I then issued a written request in the following letter:

December 12, 2003
Office of Inspector General
NASA Headquarters
Inspector General Robert W. Cobb
Sir,

It is requested that the NASA Office of Inspector General investigate the possibilities that NASA's human space flight management has been negligent in their responsibilities to provide safe and cost efficient operation of the space shuttle.

There is compelling evidence that the Columbia crew could have survived if they had been provided physiological and environmental protection (CAIB report Vol. V Appendix G.12). There is also concern that CAIB Recommendation D.a-12 Crew Survivability is not a high priority for shuttle return to flight.

There is the possibility that NASA human space flight management has impeded the automation of space shuttle operations. It is conceivable that automated flight would have significantly reduced flight costs and provided weight margins for crew escape pods.

Administrator O'Keefe testified before Congress (Feb. 2003) that he had been advised that crew escape was not feasible and that a thorough investigation had been conducted. **At this time no evidence has been produced that a thorough investigation for an automated shuttle with crew escape pods was ever conducted.**

Don A. Nelson
Retired NASA Aerospace Engineer

Mr. Cobb did respond to my letter and we had a telephone conversation in which he admitted there was no one in his office that had the technical expertise to evaluate the escape pod issue. He said the issue required that his office investigate, and that he would get people in his office with more technology qualifications. When this was done they would contact me for more information.

In September 2004, now 10 months from my request to investigate NASA's negligence in failing to investigate an automated shuttle with crew escape pods, I met with Mr. Cobb's newly hired technical experts. I explained to the "experts" (one of whom did not know anything about flight operation) that the automated shuttle in at least two NASA and one contractor studies had been found to have no technical issues preventing it from performing automated flight operations. I explained that in this automated configuration there was enough weight margin to install crew escape pods. I told them that the automated shuttle would reduce launch costs because most missions could be flown without a crew. All of this technical information existed within NASA, but NASA has refused to use it because of their steadfast stand that the shuttle was not safe without pilots. I again stated that NASA had never conducted this critical evaluation for crew escape even though Administrator O'Keefe has assured Congress that *"We will go back and look at it again, you bet. In light of this circumstance, we really do need to focus entirely on what all the alternatives are."*

Actually, this was not a technical issue; it was a failure of due diligence by NASA management to investigate every possible avenue to provide the maximum safety coverage for the crew and reduce the prohibitive launch costs. It was NASA's duty to conduct this evaluation and not merely use conjecture for their conclusion. This investigation was the NASA Inspector General's chartered responsibility and should have been investigated ten years earlier.

By April 2005 I had no indication of the status of my Inspector General investigation request. Calls to the Inspector General's office revealed Mr. Cobb's "experts" didn't work in the office anymore, and I couldn't find anyone who would answer my questions about the status of the investigation. In April 2005 I filed a Freedom of Information Act (FOIA) request on the status of the investigation that I had requested some 16 months earlier in December 2003. If the reader is thinking incompetence, I would agree. A copy of the report was finally sent to me in May 2005:

National Aeronautics and Space Administration

Office of Inspector General
Washington, DC 20546-0001
May 27 2005

Dear Mr. Nelson:

I am responding to your April 6, 2005, FOIA request that was forwarded to our office from the NASA FOIA office. You requested ". . . the findings of my request for the NASA Office of Inspector General to investigate the possibilities that NASA's human space flight management has been negligent in their responsibilities to provide safe and cost efficient operation of the Space Shuttle." My initial determination is to provide to you a copy of our Information Report titled "Crew Escape System and Automated Flight Control System - Summary of Our Actions and Conclusion."

Sincerely,

XXXXX XXXXX
Assistant Inspector General for Auditing
Office of Inspector General FOIA Officer - Audit Records
Enclosure Report Summary
In December 2003, the NASA Office of Inspector General (OIG) received a letter from retired NASA aerospace engineer Don A. Nelson. In that letter, Mr. Nelson alleged that "NASA's human space flight management has been negligent in their responsibilities to provide safe and cost efficient operation of the space shuttle."
Results
NASA's position to not pursue a Shuttle crew escape system was based on limited coverage of the flight envelope, improvements in other areas of safety, and the cost and timeline of a system. We believe that NASA's cost and timeline to implement a crew escape system was conservative and that it was highly unlikely that any schedule compression could have occurred.
In addition, Mr. Nelson could not provide us with a credible basis for his estimated cost and time to implement crew escape pods or for his projected launch cost reduction resulting from an automated flight control system.

Finally, because of the scheduled 2010 decommissioning of the Shuttle, we do not recommend that an independent study be conducted on the viability of incorporating a combined crew escape system and automated flight control system into the Shuttle at this time.

NASA had refused to "pursue" the crew escape pods which provided coverage for all phases of the flight envelope: ascent, on-orbit and entry. NASA never conducted a "cost and timeline" study. "Credibility" of cost is in the eye of the beholder...where was the NASA cost study for crew escape pods?

The NASA Office of the Inspector General failed to investigate why the automated shuttle with crew escape pods studies were not conducted in the early 1990's time frame. If they had, would this nation be buying launch services from the Russian government in 2017? Ignorance and incompetence always results in failure.

Chapter 8   Back to the Moon
You've Got to Be Kidding...

President Bush Announces New Vision for Space Exploration Program

Remarks by President George W. Bush on U.S. Space Policy
NASA Headquarters   Washington, D.C.
January 14, 2004

Excerpts:

"Inspired by all that has come before, and **guided by clear objectives,** *today we set a new course for America's space program. We will give NASA a new focus and* **vision for future exploration.** *We will* **build new ships** *to carry man forward into the universe, to gain a* **new foothold on the moon,** *and to prepare for new journeys to worlds beyond our own. To meet this goal, we will* **return the Space Shuttle to flight** *as soon as possible,* **consistent with safety concerns** *and the recommendations of the Columbia Accident Investigation Board. The Shuttle's chief purpose over the next several years will be to help finish assembly of the International Space Station. In 2010, the* **Space Shuttle -- after nearly 30 years of duty -- will be retired from service.***

Our second goal is to **develop and test a new spacecraft, the Crew Exploration Vehicle, by 2008, and to conduct the first manned mission no later than 2014.** *The Crew Exploration Vehicle will be capable of ferrying astronauts and scientists to the Space Station after the shuttle is retired. But the main* *purpose of this spacecraft will be to carry astronauts beyond our orbit to other worlds. This will be the first spacecraft of its kind since the Apollo Command Module.*

Our **third goal is to return to the moon by 2020,** *as the launching point for missions beyond. Beginning no later than 2008, we will* **send a series of robotic missions** *to the lunar surface to research and prepare for future human exploration. Using the Crew Exploration Vehicle, we will undertake* **extended human missions to the moon as early as 2015,** *with the goal of* **living and working there for increasingly extended periods.**

Returning to the moon is an important step for our space program. Establishing an extended human presence on the moon could vastly reduce the costs of further space exploration, making possible ever more ambitious missions. **Lifting heavy spacecraft and fuel out of the Earth's gravity is expensive.** *Spacecraft assembled and provisioned on the moon could escape its far lower gravity using far less energy, and thus, far less cost.*

This will be a great and unifying mission for NASA, and we know that you'll achieve it. I have directed Administrator O'Keefe to review all of NASA's current space flight and exploration activities and direct them toward the goals I have outlined. I will also **form a commission of private and public sector experts to advise on implementing the vision** *that I've outlined today. This commission will report to me within four months of its first meeting.*

Achieving these goals requires a long-term commitment. NASA's current five-year budget is $86 billion. Most of the funding we need for the new endeavors will come from reallocating $11 billion within that budget. We need some new resources, however. I will call upon Congress to **increase NASA's budget by roughly a billion dollars, spread out over the next five years.** *This increase, along with refocusing of our space agency, is a solid beginning to meet the challenges and the goals we set today. It's only a beginning."*

If I'd had been in attendance at President Bush's space policy announcement I would have shouted out in my renowned non-professional manner, "Mr. President, are you out of your cotton pickin' mind?" I and many of the aerospace engineers who had worked Apollo and the First Lunar Outpost projects knew that Bush's vision was just a rehash of those failed plans. It was conceived by a group of Washington bureaucrats who knew nothing about launch vehicle design and launch systems operations cost. The President's commission of "public sector experts" also knew nothing about launch systems development and operations. **This was not to be the only time this president was to be led astray by misinformation from so-called experts.**

While the Bush "vision of returning to the moon" in theory was feasible, it did not meet the second law for engineering requirement for success and that is, being realistic. The first and mandatory goal of NASA's next launch system must be to significantly lower launch cost. With the launch cost of these mammoth Saturn V class launchers being from $2 billion to $3 billion and budget limitations that permitted only two launches per year, there was no possibility that this limited flight schedule could support a lunar base development and long term operations. The only viable way to lower launch cost is with reusable vehicles, and Bush had just destroyed the only reusable vehicle when he ordered the decommissioning of the shuttle.

While many of the young space cadets in the aerospace community got really excited about Bush's vision, the old engineers who had worked Apollo knew that this would be another of the now many failed human space endeavors for NASA. Soon the rumors would circulate on just what would be cut from NASA's budget to pay for the Bush Vision. Center closures and layoffs were being reviewed. One major cut that did get the space science community's attention was when Administrator O'Keefe canceled the space telescope Hubble repair mission. NASA's eye to deep space would be put out to finance the Vision. Morale was already bad at NASA; Bush's vision or the lack thereof was making it worse.

Administrator O'Keefe who had no experience in developing launch systems would chose Retired Adm. Craig Steidle, who also had no experience in launch system development, to head up the Bush vision for exploration. O'Keefe would resign in 2005 to take a position in academia at reportedly three times the salary of his NASA job. Adm. Steidle would propose a fly-off competition for developing the crew exploration vehicle which would service the space station and be the crew module for the return to the moon missions. Steidle's crew vehicle launcher considerations were versions of the existing Delta IV and Atlas 5 rockets or an entirely new booster. An all-cargo only version of the Space Shuttle was rumored as being considered for support to the space station. I think this was an attempt by the so called "shuttle huggers" to keep the shuttle's orbiters out of museums. It would fail.

Adm. Steidle's plan would also fail, ending with his resignation in June 2005 after the new Administrator Mike Griffin vetoed his launch vehicle fly-off competition. NASA was now eighteen months into Bush's "new vision" and had made no progress on the shuttle replacement or the lunar launcher. Bush had been convinced to accept a four year "gap" in the time that the Space Shuttle could be replaced and was willing to depend on the Russian government during the gap to provide crew and cargo service to space station. His mandate to have the shuttle replacement in service **"no later than 2014"** was what any competent aerospace engineer knew all along; becoming impossible. The space station crew vehicle gap was still not filled in 2017.

Another Failure Looming

Even under the best of conditions, Bush's vision for returning to the Moon would have failed. We were at that time a nation with a $7.6 trillion debt and making no progress to pay it down, the September 9-11 terrorist disaster was and still is impacting our nation's budget, and both the President and NASA management ignored the unsustainable launch cost of Saturn V class launchers. Administrator Griffin would now lead the Bush Vision for human exploration, and in doing so would produce another management disaster. He dusted off his failed, First Lunar Outpost plan, which used a super Saturn V class launcher for returning to the Moon. For the shuttle replacement, the crew/cargo launcher's first stage would be the shuttle solid rocket booster and the second stage would be powered by a shuttle main engine. The configuration would be called the Ares I.

Ares Launch System

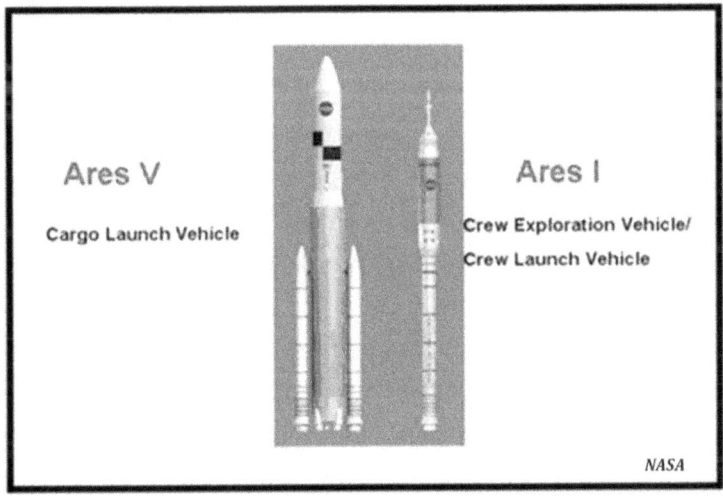

The Bush Vision plan for returning to the Moon was now to develop two expendable launch vehicles: one for crew transportation to the space station and/or the lunar transfer vehicle and another for a Saturn V class launcher for lunar missions. Then there would have to be a lunar lander/habitat vehicle that would have to be significantly larger than the Apollo lunar lander to permit the President's "extended periods of stay time" on the lunar surface. This required the lunar launch vehicle to be significantly larger than the Apollo Saturn V launcher and would require two launches to get both crew and lunar lander to the Moon.

The program would be given the catchy name, *Constellation*. The Constellation program's logo was designed to represent three development stages with first stage being human transportation to the earth's space station, then to the moon, and finally a transportation system to the planet Mars.

CONSTELLATION

To Griffin exploration team's credit, they did put together one heck of a public relation program to sell the Constellation program. The name for the shuttle replacement was Ares I and the lunar launcher Ares V. The Crew Excursion Vehicle was called the Orion and the lunar lander was the Altair.

The launchers Ares I and Ares V were to be shuttle-derived. That was another big management mistake in trying to use shuttle components for a new launch system. The shuttle main engines, external propellant tank, and solid rocket boosters that the Ares launchers were to use were designed specifically for the Space Shuttle. Even the slightest engineering change to make these shuttle components compatible with the Ares I and V expendable launchers would require an extensive verification program. Moving subsystems from one launch vehicle to another with a different configuration is never cost effective, an engineering law of economics that NASA's human space flight management still ignores. The joke around the aerospace community was the Constellation program was the Apollo program on steroids.

In this chapter we will not discuss the problems with the Ares V Moon mission launcher. Those problems will be reviewed in a later chapter on the now-planned super launcher for Mars missions. It is the Space Shuttle replacement, the Ares I, with the Orion crew/cargo modules and why it failed that this chapter will cover.

Ares I Down and Out

When I first saw the configuration of the Ares I shuttle replacement, I thought, "Oh no, another nightmare." The launcher was over 300 feet long and was given the nickname "The Stick." Any competent aerospace engineer could just look at it and see the configuration would have serious control problems. It would be like flying a wet noodle. An engineer at the Kennedy Center made this comment, *"A lot of folks here at KSC have given "The Stick" a more appropriate name ... "The Shaft" because that is exactly what NASA is doing to the taxpayers in the continued development of this unaffordable, unsustainable new launch vehicle."*

ORION CREW MODULE

The Crew Excursion Vehicle, now known as Orion' was an Apollo crew /service module configuration. Like the Apollo crew module it would have an escape rocket for first stage aborts. The crew module would also be reconfigured to carry cargo to the space station.

The Ares I/Orion configuration had big problems from the start. In the first three-year period the Orion crew module would have three different Program Managers. As the following timeline of events indicates, it was, as I expected...an engineer's nightmare.

Jul. 13, 2005: NASA awards $56 million Orion study contracts. New Ares I launcher will maximize use of Space Shuttle components which they say will have "huge cost advantages."

Jul. 14, 2005: General Accountability Office disagrees with NASA's "huge" cost saving and report cost to duplicate shuttle mission with Ares I/ Orion will be over $580 million.

Aug 10, 2005: Pentagon refuses to adopt shuttle-derived Ares I/Orion vehicle as an EELV backup because of concerns for reliability and cost of modifications.

Sept. 2005: NASA announces plan to develop methane fueled engines for Orion. NASA's research and technology program budget reduced to pay for Ares I/Orion development.

Nov. 2005: House Science Committee expresses concern for schedule and funding of exploration plan.

Jan. 2006: Orion Program Manager is removed.

Jan. 11, 2006: NASA redefines design for Ares I/ Orion because of weight and performance problems. Removes methane fueled engines requirement.

Mar. 2006: NASA is forced to extend Orion Phase 1 contracts by five months because of development problems. NASA concedes that Orion will not be ready until 2014.

In Congress, the House Science Committee had already become concerned with the progress of the shuttle replacement as the development issues became more obvious. In the following letter I tried to alert the Senate Science Committee to the Ares I/Orion (then called the *Crew Exploration Vehicle/Crew Launch Vehicle*) problems by writing Sen. Kay Bailey Hutchinson. I had been in a joint radio interview with her after the Columbia disaster and thought she might be persuaded to take action if made aware of the shuttle replacement vehicle problems. I was wrong.

April 17 2006
The Honourable Senator Kay Bailey Hutchinson
2246 Rayburn House Office Building
Washington, DC 20515-3223
Dear Senator:
As a retired NASA aerospace engineer, I and a growing number of my colleagues are very concerned about the technical and management issues with the space shuttle replacement, the Crew Exploration Vehicle/Crew Launch Vehicle (CEV/CLV). Engineering analyses are revealing that this expendable space transportation concept derived from Apollo/space shuttle components will have exorbitant development and operations cost and serious flight safety issues. Please consider the following:
* The structural and flight operation functions of this questionable configuration are in no way like those of their derived Apollo and shuttle sources. Significant and costly modifications will be required.

* *The flight control capability of the first stage is dubious. A lack of flight control authority margin and the prevailing launch winds could cause the loss of the vehicle.*

* *Launch performance shortfalls caused by errors in weight estimations are now requiring a first stage solid rocket motor with five instead of the current four segments. This configuration has never been flight tested and complicates the existing flight control problem. It also increases the probability of a Challenger type seal failure because of higher flex loads at the joints.*

* *More weight penalty is required to prevent the CEV service module reentry debris from falling into populated areas.*

* *NASA had stated that the CEV/CLV would be so safe that it would have only one catastrophic failure in 2000 flights. In fact the Apollo type parachute landing of this very heavy crew module puts the crew at extreme risk. The Russian space authorities are working to eliminate parachute landings on their next generation crewed spacecraft.*

* *The CEV/CLV cannot carry some of the space station critical components like the control moment gyros. If one of these critical components fails, the station will be lost.*

* *Because of the large manufacturing workforce needed for expendable space vehicles, the mission operations cost will exceed $700 million yearly. This budget impacting cost cannot be offset because the CEV/CLV has no commercial or military applications.*

Expendable space transportation systems have been rejected as unaffordable in two previously comprehensive studies; "Access to Space" and "First lunar Outpost". Both of these studies revealed the need for a space vehicle technology program with goals to increase flight performance, safety, and reduce operations costs. Unfortunately this has not occurred. The failure of the Evolved Expendable Launch Vehicles to attract any significant commercial interest has again confirmed that an expendable space launch system with high labor cost is not possible without costly government supplements.

NASA Administrator Michael Griffin's call for a "60 days" study to revamp the CEV concept was a warning that NASA was not in a position to support the human transportation requirements for the Vision for Space exploration. Regrettably Administrator Griffin has chosen to proceed on a course that has already proven untenable and resulted in the cancellation of the Apollo program. It is the same course he chose for the failed First Lunar Outpost project.

NASA is proceeding on this course unchallenged because there has been no independent evaluation of their "60 days" CEV/CLV. The Office of Inspector General has attempted to establish an engineering capability to evaluate space transportation concepts but failed. NASA and contractor employees who are aware of the shortcoming of the CEV/CLV are not coming forward. It is strongly recommended that the Government Accountability Office and the Congressional Budget Office be directed to evaluate cost and justification for proceeding with the CEV/CLV.

NASA Administrator Griffin has described the process of retiring the space shuttle as a "speed bump." It can more accurately be described as a "crevasse" being filled with NASA science programs for a launch concept which is already obsolete.

Don A. Nelson

Retired NASA Aerospace Engineer

Note: It is respectfully requested that this letter be made a part of the hearing records for the NASA 2007 budget.

Senator Kay Bailey Hutchison's reply:

United States Senate

June 29, 2006

Mr. Don A. Nelson

1407 Moller Rd

Alvin, TX 77511-3248

Dear Mr. Nelson:

Thank you for contacting me regarding the role of science in our nation. I welcome your thoughts and comments on this issue.

Texas is a leader in science and technology research and development and our universities and industry researchers are among the nation's best, responsible for many notable discoveries affecting our daily lives. Research and development opportunities stemming from our higher education institutions will continue to be an important driver of economic growth.

In 2004, I established the Texas Academy of Science, Engineering and Medicine to provide broader recognition of the state's top achievers. The organization, which includes members of National Academies and Texas's eleven Nobel Laureates, will bolster Texas's position as a center of achievement in these fields, identify research priorities for the future, and foster the next generation of scientists.

As a member of the Senate Commerce, Science and Transportation and Appropriations Committees, I strongly support the goals of the National Science Foundation (NSF), the National Institutes of Health (NIH) and the National Aeronautics and Space Administration, among others. These agencies play a crucial role in the support of university research in Texas, and as I continue my efforts for critical research and development funding, you may be sure I will keep your views in mind.

I appreciate hearing from you and hope you will not hesitate to keep in touch on any issue of concern to you.

Sincerely,

Kay Bailey Hutchison KBH:hcs

It is clear that the above letter is a standard auto copy reply that the Senator has sent out in response to the many letters to her office. However, even efforts to contact the Senator's staff proved fruitless. What I want to point out here is how difficult it is to get the attention of the President and Congress to address the management issues with NASA. I've made many trips to Washington to talk to members of Congress, the Executive Branch, and their staff. I've failed to convince any of them that this is not the NASA that sent men to the Moon...that this NASA's human space endeavors are failing because they have become incompetent and are not being held accountable. Well, at least I tried.

Government Accountability Office (GAO) Warning Ignored

Senator Hutchinson and her colleagues in Congress would continue to ignore the warnings that the shuttle replacement was poorly conceived. The following GAO report to Congress is one of many ignored warnings.

Excerts from July 26, 2006 report to Congress: GAO report critical of NASA CEV plans: "NASA's current acquisition strategy for the CEV places the project at risk of significant cost overruns, schedule delays, and performance shortfalls because it commits the government to a long-term product development effort before establishing a sound business case."

In following GAO reports to Congress, they would warn that the Orion crew module development cost would be nearly double the original estimate of $28 billion. Congress continued to fund this failing program.

NASA Engineers Frustrations

The following email describes the frustration of engineers at the Marshall and Johnson Centers.

Subject: Re: someone@jsc.nasa.gov

Date: Mon, 18 Jun 2007 07:

Hey Don,

Good to hear from you again.

I agree that CEV (Ares I / Orion) has problems -- BIG problems. You know how that works...we'll spend enough money to commit the nation to it and by the time the real cost comes out it will be too late and the CEV will either be finished with limited capability, or the money needed to fully complete it will be provided. You know that's how it works and so do I.

-----Original Message-----

From: xxxxxxxx

Sent: April xx 2007 10:30 PM

To: Jerry XXXXXX

Subject: (Constellation)

xxxxxx,

Constellation continues to get curiouser and curiouser. We've no sooner completed one mass reduction exercise than we were told today that another is fixing to kick-off. There's really no more mass to take off the service module w/o affecting performance. Have a feeling command module is trying to shift the load to service module.

Bottom line is we need to consider a new launch vehicle. When someone suggested the obvious — reduce crew size — the reply came back that Congress mandated into law a crew of 6 to ISS in the authorization bill, so meddling with crew size is not an option.

Over and out--

Orion XX XXXX Engineer

Reply from xxxxxx:

This is what happens when you don't have the analysis to back up a set of requirements or without a detailed reference mission.

-----Original Message-----

From: xxxx

Sent: Monday, April, 2007

To: xxxxx

Subject: (Ares I / Orion)

xxxxx, don't blame us at MSFC. We didn't dream that d____ thing up. It was a gift from two astronauts at Houston, via Headquarters, that wanted a quick (spelled safe) replacement for the Shuttle. Seems they thought it was getting too risky to fly. **And I will assure you that it did not take any great analyses to determine that we had been presented with a pig-in-a-poke.**

A five element first stage solid (that some people don't think will work), an anemic rocket engine on the upper stage that gives us a negative control authority, a lofted ascent trajectory because the underpowered J-2X cannot maintain velocity of the upper stage after separation from the first stage, and we fall 30,000 feet after separation before we burn enough fuel to change the mass fraction and let us start accelerating again.

As the old song used to say, "is that all that's bothering you, boopey?" And then, to prove boopey right, we have an upcoming Stage and a half Cargo Launch Vehicle that can't boost enough weight into orbit to allow us to go to the moon.

The Earth Departure Stage has to burn 70% of its fuel to get the lunar payload into orbit leaving 30% to perform the Lunar Injection burn, which is enough. But I asked the CaLV studiers, what do you do if the Ares / Orion can't make it to orbit as scheduled and you have to wait a month before the injection burn? All I got was blank stares and glazed eyes. I guess they figure by the time we get to the Lunar program, there will not be such things as launch delays or scrubbed launches. Go figure.

xxxx

Reply:

While I didn't know the details, I know exactly about what you speak. From what I have been reading, **you speak for many NASA engineers**

xxxxx

Aerospace Safety Advisory Panel fails to address Ares I Orion Safety

In February 2006, I addressed the NASA safety panel with my concerns about the Orion crew module. What prompted me to write the panel was what I learned at an earlier aerospace conference in Houston. The conference's main speaker was a senior spokesperson for the Constellation Program. During their question and answer session, I challenged the speaker's assumptions and pointed out there was a flight ready Saturn V vehicle sitting as a museum display not one mile from this conference room because we couldn't justify its launch cost. I asked, "How then can we now justify the launch cost of the Constellation lunar launcher?" I didn't get a reply from the speaker. What I did get was two contractor engineers working on the Constellation's Orion crew module asking if they could speak to me in private. Both said that they believed that critical safety components were being removed from the Orion crew module in order to meet the liftoff weight limitations of the Ares I. In other words the Ares I could not launch the Orion crew module because it was too heavy and management was taking out safety components to reduce the launch weight. If these contractor engineers reported this issue, they would have lost their jobs for not being "team players." I told them I would contact the NASA safety panel, but not to expect any action. My presentation to the safety panel was documented in the following report:

Aerospace Safety Advisory Panel
February 3, 2006
Marshall Space Flight Center
Huntsville, Alabama
PUBLIC STATEMENTS / COMMENTS
The first 30 minutes of the meeting were reserved for public comment on safety in NASA. Mr. Don Nelson provided a written statement to the Panel and was afforded the opportunity to make a statement at this time.

Mr. Don Nelson, a member of the public audience, introduced himself via teleconference as having a background in launch vehicles and design, and having formerly been a member of the Shuttle design team. He submitted a written statement regarding the current Exploration designs to the Panel prior to the meeting. His verbal statement paralleled his written statement and he expressed his concerns and opinion regarding the following: flight control capability; the introduction of a fifth segment creating a very top-heavy vehicle; concern about the stack related to the growing Crew Exploration Vehicle (CEV) (Orion)weight; and, perceived failure rates regarding this type of design and application.

Mr. John Frost asked if Mr. Nelson was concerned that the failure rate goal was too high or poorly assessed. Mr. Nelson replied that his point was that a 1-in-2000 failure rate was not a historically viable figure. With the current CEV configuration, he felt that the rate was more like 1-in-25 or 50.

Mr. Marshall read a portion of Mr. Nelson's written statement where he concluded that the only solution for the next-generation crew vehicle was a reusable vehicle that was developed and operated by a non-government enterprise. Marshall then queried Mr. Nelson if his assessment was based on a cost/operations perspective, safety concerns, or both. Mr. Nelson stated that his opinion was based on a 1999 study that included both factors.

Dr. Crippen took Mr. Nelson's submission under consideration for a NASA response; **the ASAP is not a technical body,** *but it will certainly consider safety allegations. Dr. Crippen could not confirm or deny Mr. Nelson's assertions on technical grounds. Mr. Nelson added that he did not think the National Research Council (NRC) had done a proper evaluation of the CEV/CLV (Crew Launch*

Vehicle). **Dr. Crippen assured Mr. Nelson that the panel would consider this assertion in a NASA response.**

There was never any response to my safety concerns. In July, 2007, I would again report to the safety panel because the safety concerns had worsened and there had been no action taken. I did not think any action would be initiated by this presentation, but I wanted the problems on record. Again the reader is reminded that **the NASA safety panel** *"is not a technical body"* and does not verify the information that NASA has provided is correct. Since the safety panel members serve at the discretion of the NASA administrator, the panel members are not in a position to challenge NASA's assertions for crew safety.

July 12, 2007
Aerospace Safety Advisory Panel
Subject: Safety and Credibility of the Orion / Ares I Constellation Program

Ref.: Presentation to ASAP, Safety and Credibility of the CEV/CLV, dated February 3, 2006.
The safety and credibility of the Orion (CEV) / Ares I (CLV) shuttle replacement vehicle has deteriorated to the point that it would be an unsafe and unreliable human transportation vehicle. Engineering analyses have proven that the assertions that the Orion / Ares I launch system would be "simple, soon, and safe" are false. Simplicity of the design was lost when none of the "existing components" which the NASA Administrator said "offer us huge cost advantages" could be incorporated into the launch vehicle. It will now cost tens of billion of dollars to redesign and flight certify these components. Statement to the media that the Orion / Ares I could be ready by 2012, have proven to be nothing more than wishful conjecture. NASA now concedes that the shuttle replacement vehicle will not be operational until 2015. Decommissioning the shuttle in 2010, mandates purchasing Soyuz launch service to space station from the Russian government (predicted minimum cost $3b). This is the same government that is kicking out the private oil companies that developed their oil fields.

Just as the "simple and soon" assertions have proven false, so has the safety claim.. The limited funding for testing and the demanding "close the launch gap" development schedule is a formula for disaster. Most concerning is ATK Launch Systems structural load analyses have failed to verify the seals in the first stage segmented booster will hold. This life threatening safety issue is confirmed by the Exploration Office scheduling the first flight tests of Ares I with a shuttle four segments booster and empty dummy fifth segment mounted on top to simulate the integrity of the seals in a launch environment. These dummy $300 million flight tests will give dummy results. THE FIVE SEGMENT FIRST STAGE BOOSTER IS UNSAFE! THIS WILL BE A DISASTER WAITING TO HAPPEN!

The second stage J-2X engine has been described by a senior NASA-JSC manager as "the long pole in the tent." Continuing reports of launch performance margin problems still questions the viability of this engine (see attached email). More concerning is that NASA is now consigning the space shuttle main engine, the world's only reusable booster engine to a museum.

Serious safety issues have also been identified with the design of the Orion command and service modules. The Orion command module is reported to have stability entry problems from 40,000 to 20,000 feet. The Panel is reminded that the only Russian cosmonauts' fatalities occurred with their crew module. Hard parachute landings still plague the Soyuz system. An Orion service module safety concern which cannot be designed away is the debris impact footprint. The suborbital service module can scatter debris into air and maritime shipping lanes and populated areas. THIS IS UNACCEPTABLE!

The International Space Station control moment gyros can only be supplied by the space shuttle. Limited storage of replacement gyros on station, cancellation of the cargo version of Orion, and the failure history of the gyros creates the threat of uncontrolled entry of the station. THIS IS UNACCEPTABLE.

The Orion / Ares I configuration was dictated by NASA and was not subjected to the comprehensive assessment of a competitive evaluation. The Government Accountability Office in July 2006 reported to Congress that NASA is using an approach "that our work has shown carries the increased risk of cost and schedule overruns and decreased technical capability." This assessment is now a reality. Congress is being pressured by the powerful aerospace lobby to support their large long term contracts for this inferior and dangerous launch vehicle.

Comments and Recommendations:

- *The safety issues with the Ares I booster structural integrity, Orion service module debris impact, and space station control moment gyro failures cannot be resolved.*

- *With all its faults the current space shuttle is a superior and safer human transportation system than the Orion / Ares I. There is a safer alternative option, privatization of the space shuttle (see: www.xxxxxx.xxx).*

- *Firing of two and resignation of a third member of the NASA Advisory Council's science committee sends the message ...don't tell the Administrator anything he doesn't want to hear.*

- *The Orion / Ares is the keystone of the Constellation Program. It has failed.*

NASA's engineers have labeled this shuttle replacement vehicle as a "pig in a poke", "the shaft", and "a monster." Their admonitions have been muted by the failure of the NASA Inspector General Office to provide a trusted environment for reporting mismanagement (see attached emails). Will the ASAP remain silent? The Old Testament prophet Zephaniah said that not listening leads to avoidable tragedy. How will history record the response of the ASAP members to NASA's greatest challenge?

Don A. Nelson
Retired NASA Aerospace Engineer
1407 Moller Rd.
Alvin TX 77511

What the safety panel reported:

Public Comment
The ASAP invited public comment. Mr. Don Nelson, a retired NASA aerospace engineer with experience in Gemini, Apollo, and lunar outpost programs, responded, stating that he felt strongly that the Administrator would lose the path to the moon. Mr. Nelson commented that NASA can't go back to the moon with expendable launch vehicles- the technology is not there to do it. It is too expensive and NASA needs reusable systems. The schedule for the Ares V is completely unacceptable. The reality of what is happening to Ares I and Orion is that it is not simple, soon, or safe. NASA is

*now looking at tens of billions of dollars to redesign and test components, and the Agency does not have the money. Instead NASA is robbing science programs by going forward with a system that should be canceled. This vehicle has created a five-year gap, with Russians supplying vehicles during that time being a highly doubtful scenario. Limited funding for testing is a formula for disaster, as is starting from scratch in a too-compressed schedule. Mr. Nelson further commented that safety issues include the structural loads on the 5-stage booster that are unacceptable. The first stage exposes seals to a high torquing force, and they will fail because they were not designed for structural wind shear loads. There is already a major weight problem with the vehicle. The Shuttle's main engine, from the only reusable launch vehicle in the world, is about to be warehoused, representing a loss of years of testing and technology. Orion also has significant problems; stability problems at 20-40,000 feet are major. Since the first of the year, reports indicate the problems have worsened. The thrusters NASA has today do not satisfy requirements. The service module will also have a debris pattern in populated areas. There has been no competitive evaluation of the Ares/Orion design; Congress is being pressured by the aerospace lobby to build an inferior and dangerous launch vehicle. A safer alternative is the privatization of Shuttle. **A trusted environment for reporting mismanagement is lacking.** Not listening leads to avoidable accidents. This concluded Mr. Nelson's remarks.*

In response, the ASAP thanked Mr. Nelson for his comments, noting that the ASAP has been sensitized to the areas of concern posed by him and will include his comments in their report.

Again, the NASA Aerospace Safety Advisory Panel is not a technical body. **Unless some "rogue" engineer like me addresses the panel, the panel knowns only what NASA is willing to share with them.** So the safety panel's reports in general only report what they have been told by NASA.

Orion Crew Module is a Death Trap

Orion Air Cushion's Failure Forces Water Recovery

"Please don't put us back in crew modules" is the plea I remember from an astronaut who had been a crew member on a Skylab space station mission. He recalled the crew training for water recovery from the command module and said that scared him more than anything else on his mission. He had a right to dread the parachute landing into uncertain sea conditions, and the recovery attempt by a helicopter's hoist. In another program there had been a drowning in a failed helicopter recovery. From landing on a runway in the orbiter to being fished out of the ocean was definitely not a safety improvement.

However, now NASA management was proclaiming that the Orion would be safer than any other human space transportation vehicle with a prediction of only one catastrophic failure in two thousand flights. When challenged, they backed off to a "maybe" one loss of crew in four hundred. Further investigation of the flight system indicates that the Orion (in my opinion) will be the most dangerous vehicle ever designed for human space transportation. Consider the following:

- Although parachutes landings have been conducted for years they still are considered very dangerous. The Russian Soyuz has recently had several near fatal incidents during the entry phase.

- The Orion crew module is the heaviest vehicle ever considered for parachute landings. The Orion weighs over one and one half that of the Apollo command module and will require water landings. A parachute deploy failure like the one on Apollo 15 will expose the crew to potential serious injury on impact.

- The Orion escape launch system is very complicated because of the heavy crew module. A crew escape tower system on a Soyuz launch failed, and there is no way to test flight the escape system. It works the first time or the crew dies.

Why the Ares I / Orion Failed

There are a number of technical and management failure issues as to why the Ares I / Orion launch system failed. These will be discussed along with the Ares V problems in the following chapters. However, the reader is asked to recall that the primary goal of the Ares I/Orion was to provide a safe and affordable replacement for the Space Shuttle. **What NASA failed to do was conduct a comparison study of the two launch systems. The agency made the incorrect assumption that the shuttle configuration was fatally flawed and could not be corrected. This failure resulted in trying to replace the Space Shuttle with a launch system that was no safer or affordable than the shuttle which President Bush had ordered to be decommissioned. Did we learn from this failed experience? What the following chapters will disclose is that we didn't. Now it's on to more failures.**

Chapter 9   The Space Launch System Debacle

President Obama Announces Human Mission to Mars
Kennedy Space Center, Florida
April 15, 2010

Excerpts:

"We will **ramp up robotic exploration of the solar system**, *including a probe of the Sun's atmosphere; new scouting missions to Mars and other destinations; and an advanced telescope to follow Hubble, allowing us to peer deeper into the universe than ever before."*

"We will increase Earth-based observation to improve our understanding of our climate and our world."

"We will **extend the life of the International Space Station;** *likely by more than five years."*

"And in order to reach the space station, we will work with a growing array of private companies competing to make getting to space easier and more affordable... By buying the services of space transportation -- rather than the vehicles themselves -- we can continue to ensure rigorous safety standards are met. But we will also accelerate the pace of innovations as companies -- from young startups to established leaders -- compete to design and build and launch new means of carrying people and materials out of our atmosphere."

"we will build on the good work already done on the Orion crew capsule. I've directed Charlie Bolden (new NASA Administrator) to immediately begin developing a rescue vehicle using this technology, so we are not forced to rely on foreign providers if it becomes necessary to quickly bring our people home from the International Space Station. And this Orion effort will be part of the technological foundation for advanced spacecraft to be used in future deep space missions.

"Next, we will invest more than $3 billion to conduct research on an advanced "heavy lift rocket" -- a vehicle to efficiently send into orbit the crew capsules, propulsion systems, and large quantities of supplies needed to reach deep space. In developing this new vehicle, we will not only look at revising or modifying older models; we want to look at new designs, new materials, new technologies that will transform not just where we can go but what we can do when we get there. And we will finalize a rocket design no later than 2015 and then begin to build it."

"At the same time, after decades of neglect, we will increase investment -- right away -- in other groundbreaking technologies that will allow astronauts to reach space sooner and more often, to travel farther and faster for less cost, and to live and work in space for longer periods of time more safely."

"Now, yes, pursuing this new strategy will require that we revise the old strategy. In part, this is because the old strategy -- including the Constellation program -- was not fulfilling its promise in many ways. That's not just my assessment; that's also the assessment of a panel of respected non-partisan experts charged with looking at these issues closely.

"By the mid-2030s, I believe we can send humans to orbit Mars and return them safely to Earth. And a landing on Mars will follow. And I expect to be around to see it."

Again, as with President Bush's Constellation announcement speech, if I had been in the audience I would have shouted out in my renowned non-professional manner, "Mr. President, are you out of your cotton pickin' mind." This was a sad day for any hope for reviving this nation's human space endeavors. President Obama had made the same mistake as President Bush did in his Constellation program announcement; he failed to understand that the cost of launch operations must be reduced in order to have a viable and affordable space program and have sufficient funding for NASA's other programs. The President had played politics in canceling the Constellation program. He appeased those members of Congress and the aerospace lobby who profit from the development and operation of the Orion crew module and development of an expendable "heavy lift rocket" by continuing those unaffordable vehicles for a Mars program. It is ironic that he would acknowledge that private companies could get to space "easier and more affordable" than NASA, but then directed NASA to have a plan for a "heavy lift rocket" no later than 2015. While NASA is trying to develop this "rocket" they will also ramp up robotic exploration, increase Earth-based observation, extend the life of the International Space Station, prepare to buy launch services from private companies, invest in groundbreaking technologies, continue to develop the over budget and unsafe Orion crew module, and plan for the human Mars program. Oh yes, don't forget to fly the Space Shuttle and plan for its decommissioning. On more than one occasion I heard the NASA Administrator Charlie Bolden confess that he was frustrated. Wonder why that was? We now have a failed human lunar program and we're on a path of certain failure for President Obama's Mars program.

Note: At Sunset High School in Dallas, Texas I had a history teacher who said that the reason we study history is because history repeats itself; if we don't learn from our failures, we will certainly fail again.

Augustine Report

President Obama's space policy was "sort of" based on the *"Review of United States Human Space Flight Plans Committee",* commonly call the Augustine report after the chairman Norm Augustine. The report was more a shopping list of options and not a defined approach for human space exploration. Although the report didn't call for the outright cancellation of Constellation, there were those in the Obama Administration who concluded that the Constellation program was so far behind schedule, underfunded, and over budget that meeting any of its goals would not be possible. And there were those who supported Constellation who said the report didn't call for cancellation. The President had the last word on the decision and Constellation was canceled.

Key statement in the Augustine report:

- *"The Committee strongly believes it is time for NASA to reassume its crucial role of developing new technologies for space."*

- *"There is now a burgeoning commercial space industry. If we craft a space architecture to provide opportunities to this industry, there is the potential—not without risk—that the costs to the government would be reduced."*

- *"The planned commercial resupply capability will be crucial to both ISS operations and utilization, it may be prudent to strengthen the incentives to the commercial providers to meet the schedule milestones."*

- *"The Committee estimates that, under the current plan, this (launch availability gap) will be at least seven years." The Committee found the interim reliance on international crew services acceptable.*

- *"If international partners are actively engaged, including on the "critical path" to success, there could be substantial benefits to foreign relations and more overall resources could become available to the human spaceflight program."*

- *"A heavy-lift launch capability to low-Earth orbit, combined with the ability to inject heavy payloads away from the Earth, is beneficial to exploration."*

- *"Mars is the ultimate destination for human exploration; but it is not the best first destination."*

There are a lot of "if's" and "but's" in the report. They accepted the launch gap when they didn't recommend extending the year the shuttle was to be decommissioned. They failed to recognize the prohibitive development and operation costs of a deep space heavy lift launcher. What the committee also failed to realize is that this nation does not have the monetary resources to conduct another Apollo concept program for human exploration. **Again, we must adopt a space transportation system which**

significantly reduces operation cost or we will forever be stuck in low earth orbit with our expendable launch vehicles.

Warning Ignored

I had tried to warn the Augustine Committee of the peril of shutting down the shuttle before an affordable and safe launch system was available. The following letter was a warning that we were losing the capability to manufacture the shuttle external tank:

May 13, 2009
Executive Office of the President
Office of Science and Technology Policy
Attn: Mr. Norman Augustine

Subject: Presidential Directive to Save Shuttle External Tanks

Mr. Augustine:
The Space Shuttle external tanks assembly line teardown is continuing despite President Obama's ordered review of the Ares Orion human transportation system (see attached email). Without the capability to make the shuttle external tanks we will be committed to purchasing foreign launch service for at least five years for space station support.
As a retired NASA aerospace advance launch system evaluator, it is my professional judgment that the space station is at extreme risk without shuttle support. ***Furthermore, NASA management disregarded recommendations that would enable the space shuttle to be a cost effective and safe 21ˢᵗ century transportation system. I would like the opportunity to discuss these recommendations with your committee.***
Regardless, the external tank assembly facility must be protected. NASA's announced schedule slip of the first crewed Orion space station mission to March 2015 and the uncertainty that the Ares I Orion management can solve its many development issues does not warrant the continued dismantling of the shuttle external tanks assembly facility. This crisis requires the immediate attention of the President.

Don A. Nelson
Retired NASA Aerospace Engineer
CC: Dr. John Holdren
        Director, Office of Science and Technology Policy

There was no reply to this letter or to the following letter in which I requested the opportunity to present the Space Shuttle Privatization Option.

June 14, 2009
To: Review of U.S. Human Space Flight Plans Committee (Augustine Committee)
Attn: Mr. Norman Augustine, Committee Chairman

Subject: Request to Present Space Shuttle Privatization Option

Space Shuttle privatization is the only option for continued U.S. human space flight that meets the essential proven requirements of being both feasible and realistic. **The commercial space shuttle (CSS) provides continued human access to space, saves thousands of aerospace jobs, avoids space station dependence on foreign launch providers, and establishes the foundation for a competitive 21st commercial launch provider with a support system for space based operations. The nation's dire economic challenge dictates that NASA must address its excessive human space flight operation costs and conduct projects with technology potential to stimulate the economy. The CSS is the most viable option to meet these requirements.**

Launch operation cost analyses indicate that the existing space shuttle is a significantly more cost effective human and cargo launch system than the heavy lift evolved expendable launch vehicle (EELV)/Orion and the Ares I Orion. Shuttle privatization establishes opportunities to further reduce operation cost of the CSS to competitive commercial launch cost, a market this nation has forfeited to foreign launch providers.

The CSS operation addresses the shuttle crew safety issue by providing crew escape pods and on orbit repair of the re-entry thermal protection system.

In response to the 107th Congress's request to investigate privatizing the space shuttle program (SSP), NASA issued the following: "Privatization of the SSP has the potential to provide significant benefits to the Government. However, timing is critical. The continuing erosion of NASA skills and experience threatens the safety of the program." Regrettably the Columbia shuttle disaster side tracked the privatization and created this unprecedented crisis for U.S. human space flight endeavours.

NASA will be required to provide technical and management support to the CSS. Unfortunately NASA is technologically bankrupt and the management of the human space programs suffers from inexperience and failure of accountability. NASA has never solved their "One NASA" problem and the centers continue to have serious disconnects that have contributed cost overruns and schedule slippages...as is the case for the Ares Orion development. The shuttle privatization presentation addresses these management issues.

The CSS option has been presented to several members of congress and/or their staff and other government space related agencies. Many suggested the presentation should be considered by the Augustine Committee.

Don A. Nelson

Aerospace Consultant

Retired NASA Aerospace Engineer

I talked to Chairman Augustine at the committee's public meeting in League City, Texas, on July 28, 2009, and asked if I could present the "Space Shuttle privatization" option. He said I would have to get the consent of their sub group Shuttle/Space Station chairperson, a former astronaut. The chairperson refused to listen to my presentation. The Space Shuttle orbiters were headed to museums and our human space endeavours were about to be set back decades.

Space Launch System (SLS)

There would be no fancy name for the replacement of the Constellation program. It would be known simply as the Space Launch System and commonly referred to as the SLS. However, some would say the SLS stood for "Senate Launch System" because after the cancellation of the

Constellation program, Congress decided that it would get in the launch vehicle design business by writing into law the design capability requirements of the SLS. The 111th Congress Public Law 267 states:

Minimum Capability Requirements.--

In general.--The Space Launch System developed pursuant to subsection (b) shall be designed to have, at a minimum, the following:

(A) The initial capability of the core elements, without an upper stage, of lifting payloads weighing between 70 tons and 100 tons into low-Earth orbit in preparation for transit for missions beyond low-Earth orbit.

(B) The capability to carry an integrated upper Earth departure stage bringing the total lift capability of the Space Launch System to 130 tons or more.

Being dutiful public servants, this is what NASA management has defined as the congressionally mandated SLS configurations:

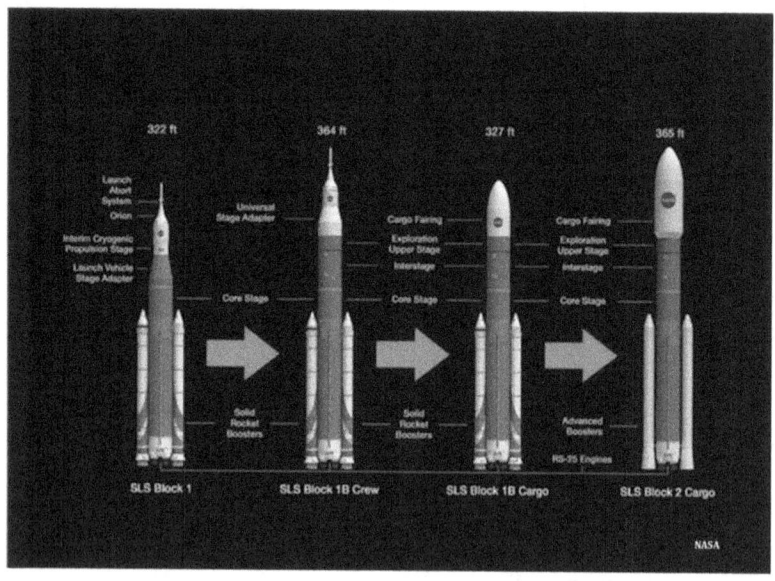

SLS Block 1 will meet the mandated 70 metric tons into low earth orbit and the SLS Block 1B will have a another design for the upper stage, giving it a 105 metric tons low earth orbit insertion capability. The propulsion system to get the 130 metric tons in orbit for the SLS Block 2 Cargo vehicle is yet to be defined. In my opinion, I don't believe the Block 1B or Block 2 Cargo will ever be built. The reason being is that once operational costs for the Block 1 are known, the program will be canceled. NASA management has been silent on the operations cost, but I've seen several independent SLS operations cost analyses including one I conducted (following) for two SLS Block 1 launches per year, the operation cost was over $4 billion.

SLS/Orion (MPCV) Operational Launch Cost Estimate
FY 2010

Manned Launch	$ mil
Flight/Launch Contract Labor Cost (11,000 total workforce JSC,MSFC,KSC)	1375
SRB (2) 5 segment	160
Civil Service	250
1st stage tanks	75
1st stage engines (5 RS-25E)	200
2nd stage (tank)	10
2nd stage engine (3 J2-X engines)	70
GFE	30
Logistic	87
Misc. (fuel,range,etc.)	77
Orion MPCV	450
Total Manned Launch	2,784

Cargo Launch

SLS manned cost less flight/launch labor fixed cost and MPCV	709
Lander (cargo)	730
Total Cargo Launch	1,439

Annual Operation Cost ~ Two Flights/YR

Total	4,223

This very conservative launch cost estimate for the Block 1 launchers is over 20 percent of NASA total budget for missions that have yet to be defined. If NASA elects to fly two SLS missions per year, the funding need to develop the propulsion systems for the Block 2 Cargo launcher would impact other NASA programs...as was the case for the Constellation program. A crewed mission to the Moon or Mars will require at the very minimum a lander, deep space habitat with radiation protection, and a propulsion module for deep space maneuvers. This nation just doesn't have the money to waste on this ill-defined and non-productive program.

Note: NASA studies have shown that a Mars human mission with a stay time of one year could require 15 launches of the SLS heavy cargo launcher...a launch vehicle which may never exist!

SLS/Orion Crew Module...Still a Death Trap

I've lost track on how many times the design of the Orion crew module has changed or how many program managers have left the program. The reader is reminded that the Constellation Orion was initially to be the shuttle replacement vehicle for getting crews to and from the Space Station. That failed, and now the SLS/Orion crew module continues to have budget and schedule problems. The crew module has been an engineer's nightmare because its weight was too heavy for the parachutes. It only has a 20 day lifetime for crew support and therefore must also have a crew habitat module for deep space missions. The limited cargo carrying capability prevents any commercial activities. Its sole purpose is to get the crew to and from earth orbit, but they have to carry it to deep space and back. For this reason alone the design should be scrapped. NASA would carry 7 metric tons of mass to deep space and back and then dump it in the ocean. What a waste!

Because the Orion crew module is so much heavier than the Apollo's, the escape tower for launch aborts is extremely complicated and therefore subject to even more manufacturing errors. The main safety problems are no protection for the crew from a failure of the pressure vessel if struck by space debris and no way to escape if any critical system fails during entry. Both problems are known safety concerns that NASA human space flight management is willing to let the crew face. These safety problems go away with crew escape pods on a reusable shuttle launch system.

NASA Administrator Charles Frank Bolden Jr.

I first met Administrator Charles Bolden when he came to the Johnson Center for astronaut training in 1980, and we've been on a first name basis since. I supported two of Charlie's four shuttle missions. In my opinion

Charlie was a top notch shuttle pilot and commander. He is a very likable, "an A number one nice guy" but today's NASA can't afford a nice guy administrator. The agency needs a butt kicker and that's not Charlie. Also, Charlie's background as a Marine pilot and astronaut made him ill prepared to head a dysfunctional governmental agency involved in very technical and scientific issues. He therefore must trust the advice of old line civil servants with personal agendas and who are known to tell the "politically appointed" administrator only what they want him to hear. Couple the administrative problems with the unsustainable Obama space policy, and it is no surprise that we've heard Charlie say that he is frustrated.

In November 2009, I wrote him the following letter. It was a failed attempt to get the Obama administration to address why NASA management was refusing to consider the privatization of the shuttle transportation system.

November 4, 2009
NASA Headquarters
Office of the Administrator
Attn: Gen. Charles F. Bolden Jr.
Subject: Revitalizing NASA
Charlie

*President Obama must address NASA's management problem. Privatization of the space shuttle transportation system eliminates the space station supply gap, insures crew access to station, saves thousands of skilled aerospace jobs, eliminates the need to develop a Saturn V class launch system, supports the commercial space initiative, reduces NASA's human space launch cost, and provides continuity for a deep space transportation system. **However, previous NASA administrations failed to conduct a comparison evaluation of the space shuttle transportation system with those being developed for the Constellation Program and the Commercial Orbital Transportation Services (COTS).** NASA's failure to conduct even a cursory comparison has led to the erroneous decision to decommission the space shuttle for what is proving to be inferior and costly launch systems. Furthermore, an independent shuttle privatization evaluation presented to the Government Accountability Office, Congressional Budget Office, and members of congress and their staff concludes that if the space shuttle transportation system is* *privatized and feasible and realistic upgrades are made, the space shuttle system will provide the safest and significantly more cost effective LEO cargo and crew transportation system.*

This failure of due diligence has misdirected the Augustine committee by not providing the advantages of a privatized space shuttle transportation system for their consideration. In addition **NASA's decision to proceed with the dismantling of the shuttle external tanks assembly without regards to the viability of the Ares I Orion shuttle replacement further implies that due diligence has not been taken.**

If this Administration fails to correct this grievous failure of due diligence it is a forgone conclusion that the space shuttle being developed by the China National Space Administration (CNSA) will propel that nation into the leadership of space faring nations.

Don A. Nelson

Nelson Aerospace Consulting

Retired NASA Aerospace Engineer

CC: XXXXX XXXX Office of Science and Technology Policy

There was no reply to this letter from NASA and the Obama administration ignored what I believe was a failure of due diligence in regards to the looming launch gap to space station, and the possibility that China would develop a space shuttle.

There were hall rumours around NASA and in the aerospace community that Headquarters had asked the Centers to consider a Plan B for human space transportation. To test that possibility, I again wrote Administrator Bolden with a twenty dollar bet challenge for the consideration of a Commercial Space Shuttle.

March 5, 2010

NASA Headquarters

Office of the Administrator

Attn: Gen. Charles F. Bolden Jr.

Subject: $20 "Plan B" Bet

Charlie,

The Commercial Space Shuttle is the only option for the "Plan B" manned spacecraft heavy-lift launch vehicle compromise. It solves the launch gap problem, has the least development cost, is commercially operated, provides the highest level of safety with crew escape pods, has proven heavy lift capability, provides the only heavy payload return capability, and has the lowest mission operation cost.

History will record that NASA management ignored requests to evaluate the commercial space shuttle as the lunar heavy lift vehicle before proceeding with their disastrous Ares Orion launch system. The hand writing is already on the wall that the Chinese space program will have a reusable space transportation system and therefore must have a space shuttle. Will history record that it was on your watch that this nation conceded human space exploration to China?

Gave the attached supporting data to Mike Coats at JSC, bet you $20 that it will never reach your desk unless Mike sends it to you.

Don

Don A. Nelson

Nelson Aerospace Consulting

Retried NASA Aerospace Engineer

I guess that I lost the twenty dollar bet that Administrator Bolden would never see my letter because I did get a reply from NASA headquarters. However, Charlie never asked for the twenty. The reply letter was written by a senior NASA headquarters public relations officer with a fancy technical title, and I'm not sure how their replied "observations and comments" were derived. One thing is certain; wherever these conclusions came from-- they were erroneous.

National Aeronautics and Space Administration

Headquarters

Washington, DC 20546-0001

April 1, 2010

Space Operations Mission Directorate

Mr. Don A. Nelson
Nelson Aerospace Consulting
1407 Moller Road Alvin, TX 77511
Dear Mr. Nelson:

Thank you for your recent letter to our National Aeronautics and Space Administration (NASA) Administrator, Charlie Bolden, proposing that NASA consider developing a Commercial Space Shuttle as the only Plan B option for a heavy-lift launch vehicle. After discussing your proposal and enclosures with Mr. Bolden, I have a few observations and comments to share.

Although **NASA does not have a Plan B,** *we are actively developing the technology, tools, and safety enhancements to make a future mission to Mars both realistic and achievable. Key to that effort will be a reliable heavy-lift propulsion system. Your letter suggests that lower operations costs can be achieved by turning the Space Shuttle over to a commercial entity. A sound business case, however, would be highly dependent on market demand beyond potential NASA requirements. Various studies and surveys, such as the annual commercial space transportation market forecast published by the Federal Aviation Administration Office of Commercial Space, indicate that there is not projected growth in the launch market. For the United States (U.S.) launch vehicle services, the market is primarily for U.S. Government payloads. Consequently, there may not be the market demand to profitably sustain a commercially-operated Space Shuttle. NASA's Fiscal Year 2011 budget request includes funding for a broad scope of Research and Development (R&D) activities aimed at developing next-generation space launch propulsion technologies. These activities aim to both reduce costs and shorten development timeframes for future heavy-lift systems. More specifically the R&D will target new approaches to first-stage launch propulsion, in-space advanced engine technology development and demonstrations, and foundational or basic propulsion research.*

Hopefully this brief explanation gives a little more insight into understanding NASA's forward plan. Thank you for your continued dedication to human spaceflight and crewed vehicle safety enhancements. These are noteworthy contributions and your comments are greatly appreciated.

XXX XXXXX for
William H. Gerstenmaier
Associate Director for Space Operations

My reply to Charlie Bolden:

Email dated April 8, 2010
Charlie:
XXX XXXXX's reply to the Commercial Space Shuttle option typifies the disconnect between NASA's senior management decisions based on conjecture and decisions made on unbiased engineering analyses.
In her letter she states:
1) "There may not be the market demand to profitably sustain a commercially-operated Space Shuttle."
The United States had lost its profitable commercial space launch market before the downturn in the satellite launch business. Our space launch industry is dependent on government support for its survival. However, while significant efforts were made to lower the operation cost of the privatized expendable launch vehicles (EELV's), none were made for the reusable space shuttle even though a NASA JSC study report that: "Privatization of the SSP has the potential to provide significant benefits to the Government. (Ref.: "Concept of Privatization of the Space Shuttle", Space Shuttle Program Office, Sept. 28, 2001).
2) "The R&D will target new approaches to first-stage launch propulsion, in-space advanced engine technology development and demonstrations, and foundational or basic propulsion research."
Exhaustive launch propulsion systems evaluations have proven again and again that no significant improvements in the performance of first stage chemical engines can be achieved. In other words, we're stuck with what we got.
In-space advance engine development can best be conducted if the engine can be tested in space and returned for evaluation. Only the space shuttle has heavy cargo return capability.
3) "NASA does not have a Plan B."
While there may be no Plan B, NASA is investigating a heavy launch vehicle (HLV) space transportation solution for the failed Ares Orion launch system. The HLV's will cost a minimum $11 billion to develop, has no commercial applications, has no cargo return

capability, and fails to solve the launch gap. The HLV like the Constellation program requires two launches to lift 66 MT to LEO for a seven day manned lunar mission.

The existing space shuttle and EELV space transportation systems can deliver the same cargo mass to LEO and avoid the launch gap and loss of thousands of shuttle jobs. Using the space shuttle and EELV eliminates the HLV development risk and cost. A commercial space shuttle further reduces operations cost. The commercial space shuttle and EELV are the better candidates for establishing a human space based transportation system for lunar, deep space, and Mars missions. A space based transportation system is mandatory for human space exploration. To continue on the Apollo expendable vehicle concept path invites failure and disaster.

In addition there is a safety issue associated in transporting astronauts to and from LEO in space capsules that has not been addressed. The Soyuz capsule has experience two fatal incidences. Warning signs of another catastrophic Soyuz capsule failure are becoming increasingly alarming. The commercial space shuttle not only significantly lowers the cost of mission operation, it can provide crew escape pods. Has NASA forgotten that crew safety is their number one priority?

I strongly recommend that NASA have an unbiased external evaluation of the commercial space shuttle and EELV space transportation system…due diligence is mandatory in this nation's critical stage of human space exploration.

Once again Charlie…this is happening on your watch.

Don

Don A. Nelson

Nelson Aerospace Consulting

Retired NASA Aerospace Engineer

Should NASA Administrator Bolden Resign?

The following letter to Administrator Bolden was an attempt to bring to his attention the consequences of continuing the SLS program. The NASA safety panel had finally admitted the SLS/Orion human launch system was **"not significantly safer than the actual historical performance of the Space Shuttle"** (Discussed in following chapters). Also the Office of the NASA Inspector General would after many ignored requests concede that they could find "no records" that NASA space flight management had conducted any evaluations that any of their shuttle replacement vehicles would be " *safer and more cost effective than the spaceflight transportation system using the existing space shuttle vehicles.*"

THE DECISION TO DECOMMISSION THE SPACE SHUTTLE TRANSPORTATION SYSTEM AND SEND THE ORBITERS TO MUSEUM WAS BASED SOLELY ON CONJECTURE.

Letter requesting Administrator Bolden to resign:

February 23, 2015
NASA Headquarters
Office of the Administrator
Gen. Charles F. Bolden Jr
Subject: Resignation
Charlie:

The NASA Aerospace Safety Advisory Panel 2014 annual report states that: **"(ASAP) expresses concern that the Loss of Crew probability thresholds for them (SLS/Orion) are not significantly safer than the actual historical performance of the Space Shuttle."** *The Space Shuttle was decommissioned because it was perceived to be unaffordable and unsafe for continued manned operations. What the ASAP 2014 annual report revealed was that NASA management has mislead the Congress and the President by failing to report that the Orion's "crew module configurations" has had a dismal safety record. Instead the NASA management position has been that the Orion crew module will be ten times safer during launch and entry than the space shuttle (NASA release 11-164). Had the Congress and the President been advised by NASA that the SLS/Orion was not significantly safer than the Space Shuttle they would have undoubtedly elected to continue Space Shuttle operations to the space station rather than depend on the dubious Russian government for human assess to the station in a Soyuz "crew module" that is not significantly safer than the decommissioned space shuttle.*
Furthermore an independent commercial space shuttle freighter study (see website: spacetran21.org) conducted by concerned aerospace engineers concluded that existing

technologies are available for developing a commercial space launch system. The commercial freighter provides improved crew safety with crew escape pods, significantly lower launch costs for manned and unmanned LEO and deep space missions, and would have the capability to meet the USAF need for rapid response to foreign or deep space threats (comets/asteroids). Instead of conducting requested investigations for a commercial space shuttle, your administration elected to proceed with the decommissioning of the space shuttle and try to develop the obsolete and unsafe SLS/Orion. Looming SLS/Orion schedule slippage, dubious mission goals, questionable development and operations costs, and increasing morale problems has this nation's human space programs on the verge of collapse.

Enclosed is a paper trail which documents the Orion crew safety issues. The NASA Office of the Inspector General has been requested to determine if this misrepresentation of the Orion crew safety was an act of duplicity or management incompetence. There is no doubt that the SLS/Orion will fail. To prolong the cancellation of this failed effort will be devastating to the NASA community. Therefore it is in the best interest of NASA and the nation that you acknowledge the SLS/Orion is a failure and resigns as NASA Administrator.

Don

Don A. Nelson

Retired NASA Aerospace Engineer

Coordinator, Concerned American Aerospace Engineers

NASA Administrator Bolden's email reply sent 7-14-2011

Don,

As I hope you know, I have the utmost respect for your professional capabilities and competence. While I have no intention of submitting my resignation as you recommend, I can understand your frustration with the situation in which we find ourselves and intend to continue to work all day, every day to bring the nation an affordable and sustainable exploration program that works in synergy with our science, aeronautics, and technology development efforts.

Respectfully,

Charlie B.

Charles F. Bolden Jr.
Administrator
National Aeronautics and Space Administration
300 E St. SW
Washington, DC 20546-0001
(202) 358-XXXX

Administrator Bolden was again challenged by Chris Kraft, a former NASA Johnson Center director, who was reported to say in a *Houston Chronicle* interview that the Space Launch System rocket "will eat NASA alive." Asked about this comment in a congressional hearing, Administrator Bolden replied, *"I have the advantage of a team around me that he didn't have,"..."You have to remember. Most of us forget.* **I have a very mature leadership team**. *When Dr. Kraft was in mission control, and when he led the Johnson Space Center, we went to the Moon. Most of the people were 20 years old.* **They didn't know anything.** *"*

Administrator Bolden's **"very mature leader team"** has given us a failed Constellation Program, dependence on the Russian government for access to the space station, a failing SLS program with no mission assignments, no significant improvement in crew safety, and no possibility of providing affordable launch costs.

Letter to the NASA Inspector General:

Nelson Aerospace Consulting
1407 Moller Rd Alvin Texas, 77511
March 6, 2015

Office of Inspector General
XXXXX XXXX, NASA Inspector General
NASA Headquarters

Ref.: Letter dated February 5. 2014, The OIG is requested to investigate if the misrepresentation of Orion crew safety was an act of duplicity or management incompetence. *OIG is requested to investigate if the failure of NASA management to privatize the Space Shuttle was an act of duplicity or management incompetence.*

Mr. XXXX:

Congressman John Culberson, Chairman of the Commerce, Justice and Science (CJS) Appropriations Subcommittee in his March 4, 2015 meeting stated that: "Administrator Bolden made it clear in his answers that the Obama Administration has no contingency plan in place to send U.S. astronauts to the International Space Station if Russia chooses to end the current agreement that allows our astronauts to travel to the space station on board its Soyuz capsules." **What was not discussed in the meeting was that Administrator Bolden had failed to inform the Congress that the then existing Space Shuttles could have continued to send astronauts to the space station at no significant risk greater than that of the crew modules to be used for the SLS/Orion and NASA's Commercial Crew Programs.**

What Administrator Bolden also failed to reveal is that the then existing space shuttles could have been automated and equipped with crew escape pods making the Space Shuttle the safest crew transportation available. What Administrator Bolden did not disclose was that there is now existing technology which if incorporated would make the space shuttle concept the most affordable and safest human transportation available.

Mr. XXXX, when it comes to astronaut safety and the future of our space program you will find I am very tenacious. I encourage the IGO not to prolong this requested investigation.

Don A. Nelson
Retired NASA Aerospace Engineer
Coordinator Concerned American Aerospace Engineers

The NASA Inspector General's office ignored this request as well as the February 2014 request for NASA investigation for failure of due diligence. One would think this office works for NASA and not the U.S. taxpayer. Then the reader is reminded that a previous Inspector General admitted this office was not qualified to make technical investigations.

In the opinion of this old NASA engineer, the Space Launch System will eventually be canceled. I think Charlie Bolden inherited a dead horse...one that no one will ever ride to the stars.

Chapter 10   Commercial Cargo/Crew Program

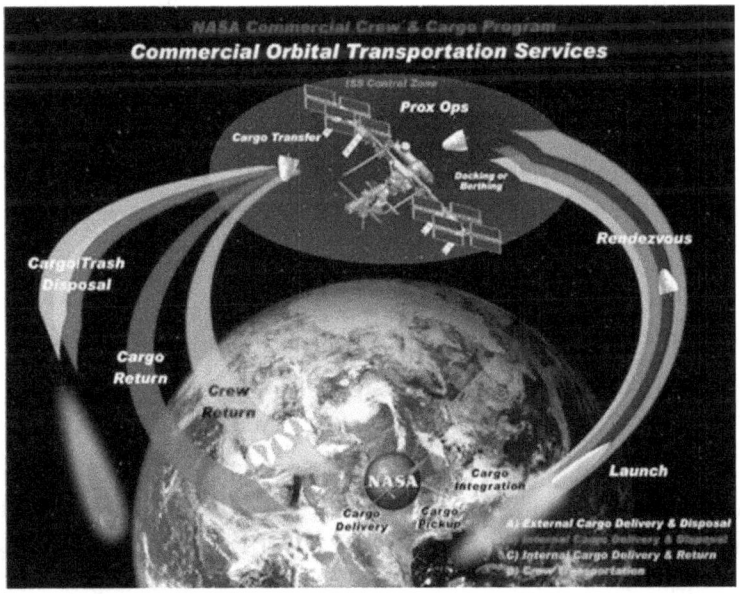

In 2016 the NASA Office of the Inspector General released a report on their investigation of the troubled Commercial Crew Program. This was a rare case where the Inspector General's office actually did their job and conducted a "somewhat" comprehensive review of a problem with NASA's human space transportation systems. The report concluded that the commercial crew vehicles would not be available for service to the space station until late 2018. The reader will recall that President Bush decommissioned the shuttle because he was led to believe the shuttle replacement would be ready "no later than 2014." The NASA Inspector General's report revealed that NASA will, by 2018, have given $3.4 billion in payments to the Russian government for transporting 64 astronauts to and from the space station. If you add in the cost of the failed Ares I/Orion program, development cost of the commercial cargo and crew vehicles, and decommissioning of the Space Shuttle, the result is that the replacement cost of the shuttle so far will have exceeded $17 billion.

While the 2016 Inspector General's report is noteworthy, it does not excuse their failure of due diligence in refusing to conduct an investigation to insure that the shuttle replacement would indeed be more affordable and safer than the Space Shuttle. Instead, in 2017, we will still be buying crew launch services from the Russians, and **if the commercial crew vehicles are ever flight qualified they will not be significantly safer than the decommissioned Space Shuttle**. In the opinion of this old NASA engineer, the NASA Office of the Inspector General is a primary contributor to the failure of this nation's human space programs.

NASA's Commercial Cargo/Crew Programs

In 2005, NASA Administrator Griffin planned for the allocation of over $500 million for a five year program for the development and demonstration of commercial space transportation for cargo and crew delivery to the space station. This was necessary because the shuttle replacement vehicle, the Ares I / Orion crew module, was designed for six crew members, making it too large and too expensive for service to the space station. The privatized Space Shuttle could have met the space station's servicing needs, but NASA's human space flight management was steadfastly opposed to continuing any type of shuttle operations. Again the reader is asked to remember that NASA management also steadfastly refused to conduct any evaluations to verify that their shuttle replacements were more affordable and safer.

There are also major concerns for the commercial orbital transportation servicing plan. The commercial cargo and crew providers only have one customer and one facility to service. If NASA decides to vacate the space station or it becomes uninhabitable, there is no market for the provider's space services. Although the current plan for the station is to remain in service until 2024 that might not be what happens. The world economy is becoming more uncertain and the probability of a critical element failing on the station makes achieving this date very doubtful.

Commercial Cargo Program

There have been some positives in NASA's commercial approach to space transportation. Two companies, SpaceX and Orbital Sciences, have been successful in providing commercial cargo service to the station. Both companies have had problems with their **expendable** launch vehicles. SpaceX has lost two Falcon 9 launchers and Orbital Sciences had a failure of their Antares launcher. Below is the SpaceX Dragon cargo vehicle being retrieved by the space station remote arm. The first flight was in December 2010, and regular cargo flights began in October 2012. The Dragon cargo vehicle is launched on the SpaceX privately developed launch vehicle the Falcon 9. It has a payload delivery and return weight of 7,300 pounds. The Orbital Sciences Cygnus's (below) first mission to space station was in September 2013. The first scheduled resupply mission arrived at the station in January 2014. It has a payload weight of 7,700 pounds and does not return any payload.

The Sierra Nevada's Dream Chaser (below) was rejected as a crew vehicle but approved to continue being developed as a potential cargo resupply vehicle. It will be launched on the Atlas V when ready for orbital flight tests and therefore making it another unaffordable launch system.

Sierra Navada's Dream Chaser

Commercial Crew Program

Providing crew transportation to the station is proving to be difficult. There are two companies, Boeing and SpaceX, struggling to develop a crew module vehicle for the station by late 2018.

Spacex Crew Dragon

Boeing Starliner

NASA

Commercial Crew Modules are Death Traps

Congress continues to fund the development of the commercial crew modules, and these modules should eventually be available for crew transportation to the space station. However, the NASA Astronaut Office has requested that the next crew transportation system have an order of magnitude safety improvement over the Space Shuttle during ascent with a predicted loss of crew of one in one thousand launches. They want to have "a full envelope abort/escape capability with no black holes." My interpretation of "a full envelope" is that the crew must have not only an ascent escape system but also an on-orbit and reentry escape capability. To meet this requirement for on-orbit and entry, there must be crew escape pods. Therefore the commercial Boeing and SpaceX crew modules cannot meet this requirement because there is not sufficient volume in either to install escape pods. The reader is reminded that one of reasons the Space

Shuttle was decommissioned was to have a launch system that improves crew safety, and that is not the case with crew modules! This is another NASA management blunder, which again will have fatal consequences!

Crew Modules Failures

Crew modules have a history of fatal occurrences, the Apollo 1 fire and two Russian Soyuz crew modules. There have also been numerous near fatal incidences. Problem areas are:

- Every crew module flight is a test flight! Manufacturing errors have and will continue to occur.
- Water landings are an unacceptable dangerous risk to flight and recovery crews. Land landings have the potential to expose the crew to fatal high-g loads and/or penetration by sharp rocks or fence posts upon impact.
- Crew modules have very limited cross range capability which could require a reentry into unacceptable weather conditions.
- Crew module's notorious reentry errors can result in an expanded landing zone that could prevent rapid access to a crew in dire circumstances.
- Parachutes are known to fail.

Orion Crew Module Crash Site

On July 31, 2008, during this drop test, a Orion's crew module parachute failed to inflate. Even on the production crew modules, where every flight will be a test flight, there is the possibility of a parachute failure or some other critical system failure. The reader is reminded of the parachute failure on the Apollo 15 mission. **NASA had considered a land landing for the Orion crew module but after this failure, this option was dropped. However, even a water landing in rough sea conditions could induce a force upon impact that could cause a serious or fatal injury to the crew. Currently, the Boeing Starliner has plans for a land landing, but the real possibility of impacting on a sharp rock or fencepost will make this option unacceptable.**

The above photo is of the failed attempt to recover the Mercury Liberty Bell 7 after it started taking on water. Note how calm the ocean is during this failed recovery operation.

NASA Crew Safety Predictions Erroneous

The NASA Director of Commercial Space Flight has been quoted as saying that the prediction of a commercial crew loss is only one in 270 flights and is three times safer than the Space Shuttle. I challenged the director to make public the analysis for their speculative safety claims in the following letter:

Nelson Aerospace Consulting

1407 Moller Rd

Alvin Texas, 77511

August 17, 2015

NASA Headquarters

Mr. XXX XXXXXXXXX

Director of Commercial Space Flight

Mr. XXXXXXXXX

Subject: CCP Loss of Crew Analyses

In the 8/13/15 issue of "The Verge" you were quoted to say : . XXXXXXXXX, director of commercial spaceflight development at NASA, said, **"For every hundred missions, how many missions could you analytically show are going to be safe and return the crew safely to Earth?"... "The number we've come up with is: for every 270 flights, we might have one where we're going to have a bad day."** Based on your statement the article's author concluded that the **commercial crew modules are theoretically three times safer than the decommissioned space shuttle.**

The ASAP in their 2014 Annual report stated **that loss of crew probabilities for crew modules like the SLS/Orion "are not significantly safer than the actual historical performance of the Space Shuttle."** A FOIA requested by this engineer found that NASA has never conducted any analyses confirming that the SLS/Orion launch vehicle and/or crew modules are safer than the Space Shuttle. **Ever crew module launched on an expendable launch vehicle is a test flight subjecting the crew to unknown manufacturing errors and has no crew escape system for entry. Whereas a reusable space shuttle flown through a flight test program will have proven flight systems reliability.** In addition a space shuttle is the only crew transportation vehicle that can provide crew escape pods for accent, on-orbit, and entry catastrophic events. Since the design of the decommissioned Space Shuttle there has been significant improvements in flight systems which if adopted in the shuttle configuration will provide a more cost effective and safer flight (see: www.spacetran21.org). However, NASA management has steadfastly refused to evaluate this flight system.

Unless you have validated analyses that the CCP crew capsules are safer than the decommissioned shuttle as your reported statement indicates you are duty bound to inform the GAO and the Congress of this grievous NASA management blunder.
Don A. Nelson

There was no reply to this letter. The NASA safety panel now states that NASA's 1 in 270 failure "statement" is a goal for the commercial crew module and not supported by analysis. **Again the NASA safety panel has covered for a NASA management failure to admit the safety issues with crew modules.** The historic shuttle loss was 1 in 67 launches and there is no way to make crew modules "significantly safer."

Chapter 11 Commercial Space Shuttle Freighter
Hey NASA Space Flight Management…Guess What's Back!

Engineers can't leave any project until they are convinced it is obsolete and another concept can replace it. Only then should it be placed in a museum or in the scrap heap. This old NASA engineer hasn't been convinced that the Space Shuttle launch concept is obsolete. It is true that there were more problems with the decommissioned shuttle than just poor management. First, it was an engineer's nightmare to upgrade the computer software and the flight subsystems. Second, the landing weight was barely was within the safety limits. Third, the thermal protection system was obsolete and proved to be deadly in the Columbia disaster. However, even with all its faults, the Space Shuttle was still a better launch system than any of the proposed launchers that have been tried and failed to replace it. In fact, there have been significant technical achievements in the last decade that when assimilated into the shuttle's flight system would make the launcher far superior to any current or future expendable launcher. This is the case for the Commercial Space Shuttle (CSS) freighter which is the result of the integration of many advanced flight systems. These advanced flight systems made the shuttle concept more reliable, affordable, and safer to operate. There is another major difference; the CSS freighter only takes payloads to and from low earth orbit. Unlike the decommissioned shuttle, it does not serve as an on-orbit space laboratory. The CSS freighter is being promoted as the replacement for NASA's failing Space Launch System.

Commercial Space Shuttle Freighter

From the outside, the CSS freighter will look like the decommissioned shuttle. With existing technology, the CSS freighter will be able to provide the following services:

CSS Freighter Design Reference Mission

- Deliver and return 20 metric tons to and from a circular orbit of 240 nm at an inclination of 28.5 degrees.
- Orbiter return to launch flight status will not exceed five days.
- All mission phases conducted by autonomous operation with ground control backup flight guidance.
- Provide escape pods for up to 4 passengers.
- Provide for 4 days on-orbit operations plus 8 days with backup power down mode when carrying passengers. Dormant (no passengers) on-orbit systems will be designed for 60 days in gravity gradient attitude for repair and retrieval of disabled freighter.

In promoting their Saturn class heavy launchers and crew modules, NASA space flight management had really done a hatchet job in discrediting the Space Shuttle launch concept. Many in the aerospace community still believe the shuttle launch concept was a failure and that NASA was justified in abandoning the flight system. In an effort to restore the credibility of the shuttle launch system, I had written articles for space publications, contacted aerospace company management, and briefed members of Congress on the superior capability of the Commercial Space Shuttle (CSS). However, now that the Space Shuttle orbiters are in museums these briefings have a different approach as they show how the Concerned American Aerospace Engineers (CAAE) have incorporated existing technology into the decommissioned Space Shuttle launch concept. Some of the information I have presented on this revised shuttle concept is in the following commentary article I wrote for a technical magazine in June 2015:

"Space Launch System versus the Commercial Space Shuttle Freighter...a Reality Check"
By Don A. Nelson

There are two options for the LEO phase of a human deep space transportation system...an expendable heavy lift Apollo class launcher with a crew module and a reusable space shuttle. As an aerospace engineer at NASA-JSC, I had the privilege of working on the design and operations of both these options. I also served as a mission operation evaluator on numerous launch system proposals including the Advance Launch System and Access to Space studies. In my professional opinion all of these studies and programs were canceled because NASA management has refused to concede the development and operation of launch systems to the commercial sector. **Launch transportation systems cannot rely solely on government support for sustainability and therefore must have commercial involvement to reduce operation costs. The flight rate must be maximized to lower the annual fixed cost (employees, facilities, suppliers, etc.) which requires a launch system that can provide launch service for civil, military, and commercial launch needs. In addition, the launch system must provide the safest human transportation possible. This goal can only be accomplished with commercially operated reusable space shuttles that support space based tugs and cruisers which provide near earth and deep space transportation.**

NASA promotes the Space Launch System (SLS) as being safe, affordable, and sustainable. How can the SLS be affordable since all previous programs with expendable heavy lift Apollo class launchers and crew modules have been canceled because they proved to be unaffordable? Neither the 70 mt or 130 mt SLS launchers has any military or commercial applications which are needed to reduce annual operation costs. **Therefore the SLS is sustainable only as long as Congress is willing to provide funding.** *Development and operations costs remain a closely guarded NASA secret because if these costs were made public even NASA management believes the "sticker shock" would kill the program. Even if the SLS operations cost were provided by Congress, the Orion's crew and service modules has unsolvable design* *flaws. The Orion's mission is that of a LEO taxis to get the crew to and from orbit. On deep space missions that are longer than the 21 day lifetime of the Orion, the service module is useless dead weight since an orbit transfer stage will be required to perform all the propulsion maneuvers to and from deep space. Also the Orion crew module has no backup escape system for re-entry. All the Orion re-entry systems must work the first time or you lose the crew. NASA's claim that the Orion will be ten times safer during entry than the decommissioned Space Shuttle was challenged by the Aerospace Safety Advisory Panel. In their 2014 annual report the panel reported that the SLS Orion was not significantly safer than the decommissioned space shuttle since there is no crew escape system for entry failures. This also means that the Commercial Crew Program's crew modules have the same entry safety issue. Only an entry crew escape pod would increase crew survivability. Because of inadequate volume in crew modules, crew escape pods cannot be installed...all critical entry systems must work or the crew dies. In reality NASA management has shown no evidence that the SLS Orion program is safer or even more affordable than the decommissioned Space Shuttle program.*

In contrast to the SLS program, a commercial space shuttle (CSS) freighter would be significantly safer and definitely more affordable. The CSS freighter is the innovation of a small number of aerospace engineers who are convinced that it is the only option which can provide safe and affordable LEO transportation. *The design reference mission is to deliver and return 20 mt to a circular orbit of 240 nm at 28.5 degree inclination, be developed with existing technologies, and have a launch turnaround capability of 5 days. The CSS freighter may look like the decommissioned Space Shuttle but its construction and operation is vastly different. The orbiter, external tank, and solid rocket booster are constructed primarily of composite materials which results in a significant weight margin and reduces manufacturing cost for the expendable tank and booster motors. The freighter is designed for maximum affordability. All subsystems are modular with a plug-in replacement capability, "green" propellants with no hydrazine, upgraded main engines, long life batteries with solar arrays, launcher assembled at launch pad, ship and shoot payloads, and crew escape pods for manned flights. The thermal protect system is 4th generations tiles which can be repaired or replaced on-orbit. In this configuration and with no civil service overhead it can provide safer and affordable support for civil, military, and* *commercial near earth and deep space endeavors. It can compete in the international launch market and has the unique capability to return payload from LEO. The Space Shuttle role in the assembly and support of the space station proves the concept for the CSS freighter constructing and supporting space tugs and deep space cruisers for near earth and deep space missions. An extremely important capability is the rapid launch turn around provides the USAF's with a solution for responding to sudden foreign or deep space (comets, asteroids) threats. However, at a USAF Space and Missile Systems Center briefing on the CSS freighter, concern was expressed that the overall flight rate would not be sufficient to lower the operations cost enough for profitable operations. I concurred with their observation and my reply was what price would be paid if we failed to meet a threat to the nation or to our world?*

The China National Space Administration is reported to be considering a space shuttle development. China's economic strategy is based on securing natural resources and as shown by their lunar program they are looking to space to provide these future needs and this will require a space shuttle. **It has been the goal of every space faring nation to develop reusable space transportation systems. However, only this nation has the existing technology to accomplish that goal. The only obstacles for a CSS freighter are political.**

Don A. Nelson
Coordinator Commercial Space Shuttle Freighter

By this time the Space Shuttle's orbiters have all been consigned to museums. The capability to continue flight operation with these still viable vehicles is forever lost. The original goal was to privatize the shuttle fleet and continue operations until the second generation shuttle fleet could be designed and built. It was recognized that the existing orbiters could not be upgraded as they existed and would have to be stripped down and refitted with modular subsystems. However, it was still the best approach to use the existing shuttles rather than accept the launch gap in this nation's human space transportation capability. Another problem was that to build new orbiters it would have required that the jigs used to form the aluminum metal into the orbiter fuselage be saved. As it has turned out the jigs will not be required to build the next generation orbiter.

Commercial Space Shuttle Freighter Configuration

An important difference between the decommissioned Space Shuttle and the CSS Freighter is that the freighter will be designed for modular subsystems. For example, if one of the onboard computers or propulsion systems should fail or need to be serviced, the maintenance service personnel would just disconnect it and "plug in" the replacement. Also, if the subsystem would have become obsolete, it could be un-plugged and the upgraded subsystem plugged-in. This is not new technology and has been used on commercial airliners and military aircraft for some time. It is a mandatory requirement for rapid return to launch capability. Other significant features of the freighter are shown in the following:

Commercial Space Shuttle Freighter Configuration

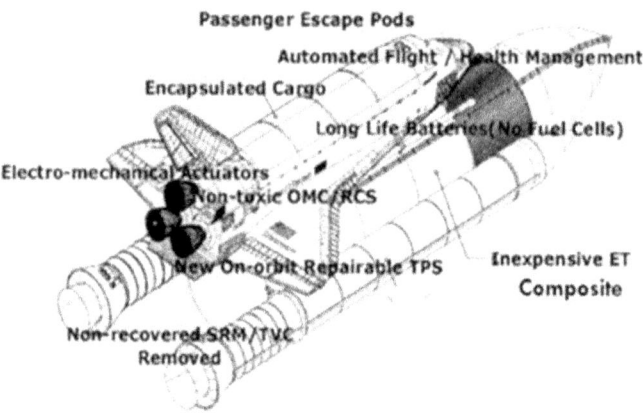

Passenger Escape Pods
Automated Flight / Health Management
Encapsulated Cargo
Long Life Batteries(No Fuel Cells)
Electro-mechanical Actuators
Non-toxic OMC/RCS
New On-orbit Repairable TPS
Inexpensive ET Composite
Non-recovered SRM/TVC Removed

Subsystems must be a modular concept, designed for unconstrained replacement. See: www.spacetran21.org

In my professional judgment, the most significant change for the CSS freighter configuration is the use of composites. Using composites in the fuselage of the orbiter, solid rocket motor casing, and external tank will significantly reduce the structural weight and give the design engineers more options for improving operations. Composites have already been used for these purposes in commercial airliners and military aircraft. Composites are being used in Sierra Nevada's Dream Chaser reusable orbital spacecraft.

The flight proven USAF X-37b uses light weight composite structures and has had a number of successful long duration orbital missions.

Crew Escape Pods

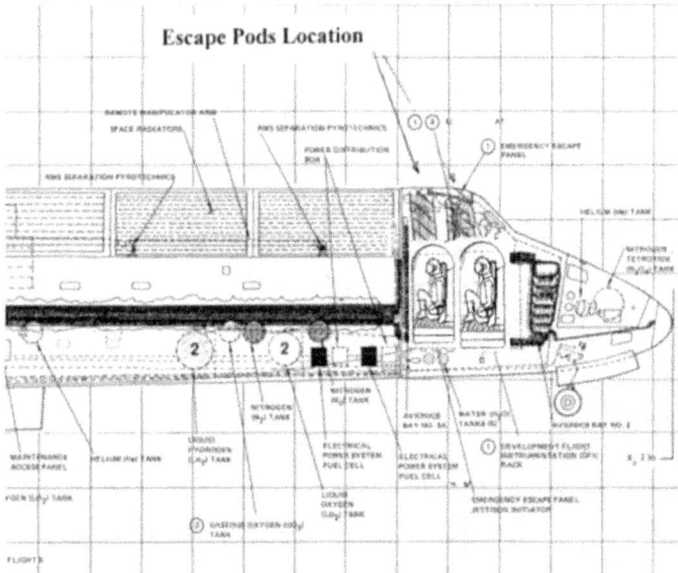

CSS Freighter/Crew Escape Pods

Removing the piloting function from the orbiter permits the installation of the crew escape pods. There are many options for the design of the pods. The design requirements are to provide escape functions for both ascent and entry. The pod will serve as a safe haven facility in the event the orbiter is unable to deorbit and a rescue mission can be launched. Also, the pod will serve as a safe haven if cabin pressure is lost. Pod design will provide the occupant with communication and life support.

CSS Freighter Fly Back Boosters

It is expected that the initial CCS freighter launch system will use the existing solid rocket boosters. However, it may prove that fly back boosters can be developed which would significantly lower launch operation cost and improve reliability.

CSS Freighter Business Model

By now the reader knows that I and a few other Concerned American Aerospace Engineers believe that a commercially operated Space Shuttle launcher concept is the only affordable and safe option for 21st century space travel. In a previous chapter, I stated the launch cost for a commercial shuttle could be as low as $205 million, which is not a competitive launch cost in today's launch market. However, this launch cost was determined using NASA operation data. The following includes ways by which the freighter's launch costs can be made competitive.

Since launch cost is a function of launch rate, the following chart is a "crystal ball" outlook of what "could" be the annual launch market for the CSS freighter:

Commercial Launch Market for CSS Freighter

SPACE SHUTTLE TO 2030
"MARKETS"

• COMMERCIAL	4 TO 16
• MILITARY	2 TO 8
• SPACE STATION	2 TO 4
• EARTH OBSERVATIONS	2 TO 4
• DEEP SPACE PROBES	1 TO 2
• LUNAR EXPLORATION	2 TO 30
TOTAL	12 TO 64

Launch cost is a function of launch rate and all launch systems require government funding.
The CSS Freighter is no exception, however it will offer superior launch service with the capability to return non-functioning payloads

The reader will note that the minimum rate of only 12 launches per year would not sustain a commercial enterprise especially since two of those launches would be for a space station that may have been decommissioned. Therefore, the CSS freighter must be subsidized by the government until commercial launches becomes the sustaining market. It is not unreasonable to expect the commercial launch market to significantly increase if the world's economy remains strong. Also, the CSS freighter offers some very unique services that the current launch companies cannot provide:

• The ability to save the customer's payload in the event of a launch failure by aborting back to the launch site and landing.

• Retrieval of the customer's payload if it develops on-orbit problems or fix it on-orbit.

• Provide the customer with more launch opportunities because of the CSS freighter's rapid launch turn around capability.

• Lower customer's insurance rates with flight-proven reusable launchers that also save the payloads.

• Provide the only capability for returning commercial resources from space mining and/or manufacturing.

- Serve as an orbital construction platform for assembly of space stations, near earth and deep space reusable spacecraft.

These unique capabilities also apply to the military and NASA's launch requirements, especially the capability to "rapidly" meet a deep space threat to earth.

The bottom line here is that the CSS freighter will be a valuable government initial investment and not just another jobs program. Unlike the NASA and USAF expendable vehicle systems, the CSS freighter has the potential of becoming self- sustainable. Therefore, the recommended financial plan is similar to that proposed for the NASA X-33/VentureStar program. A private company will develop the CSS freighter with existing technologies. More than one company may want to participate in the initial development and operational phases. The government will provide a "portion" of the developing funding as a loan. Since only existing technologies will be used, the development cost should be significantly lower. Also, the freighter will have only one mission and that is to deliver cargo and passengers to low earth orbit…again, a factor that will lower development and operations costs. When the freighter is ready for flight tests the developer may want to subcontract or lease the freighter(s) to a launch operations company. During the freighter flight test phase, which may be 50 or more missions, the government (NASA, USAF, etc.) will annually purchase enough launches to keep the company solvent. When the CSS freighter becomes profitable, the government annual funding loans will be reevaluated. **This is a low-risk low-cost approach to lowering launch cost!**

CSS Freighter Long Range Plan

As the following chart states, **"To get somewhere…we have to know where we're going!"** A lack of achievable goals has been the problem with all of NASA's human space endeavors. They all failed to gain the support of the aerospace community because their long range objectives were either unaffordable and/or unsustainable. What the "Space Based Transportation Plan" does is outline a long range plan that is both affordable and sustainable.

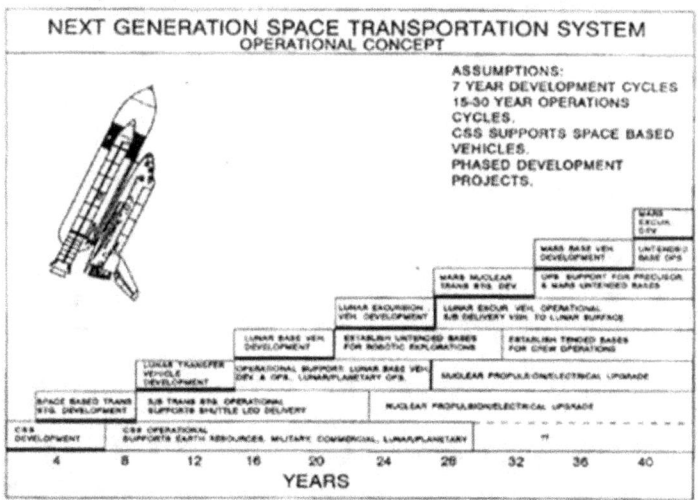

Space Based Transportation Plan
"To get somewhere... we have to know where we're going!"

With the CSS freighter as the cornerstone for the Space Based Transportation System, the space based long-range operation plan can be revised to comply with a change in the economic environment without seriously affecting the function of the commercial operations. Also, the long range goals can be changed from lunar to Mars or whatever as the space operations needs dictate.

The Marshal Center proposed space tug is a candidate for the space based vehicle:

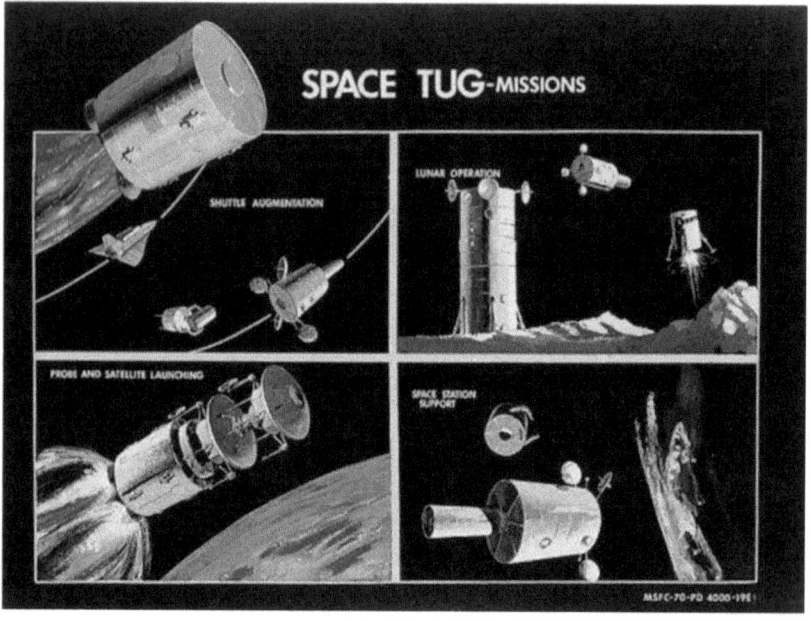

Deep Space Transportation

The first priority for robotic and human deep space exploration and potential commercial activity is to develop an affordable transportation system. The logical path is first to develop space based vehicles that are designed to transport robotic payloads that expand our knowledge of the solar system, and as a secondary objective identify avenues and designations for later human deep space travels and/or commercial utilization. The role of the CSS freighter will be to support deep space programs by constructing and supplying space based vehicles in lower earth orbit. Again the Marshal Center has a viable concept:

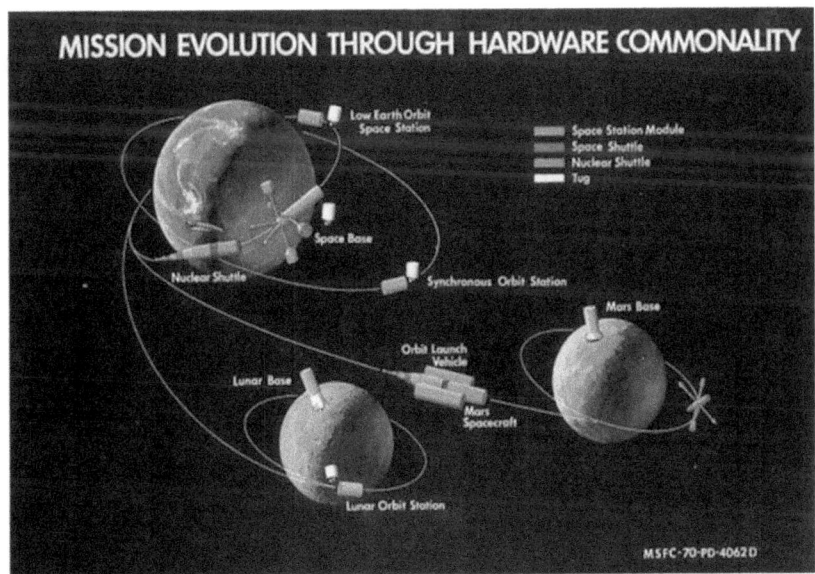

An example of a future space based vehicle for deep space transportation is the nuclear thermal rocket (NTR) as shown below:

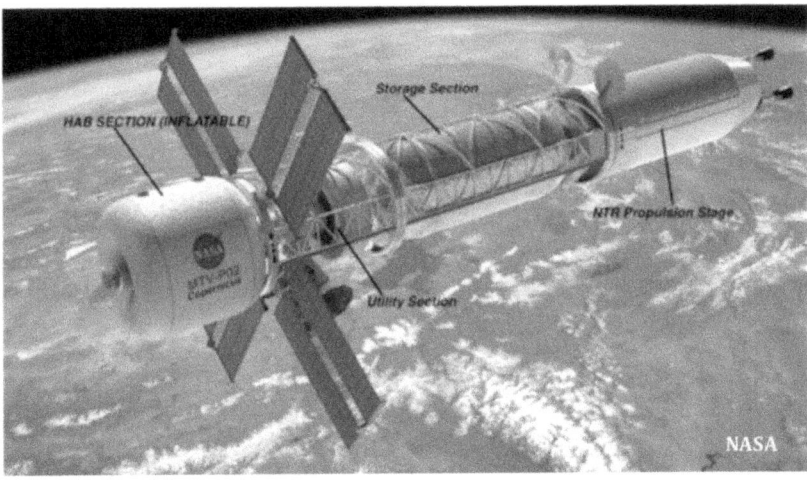

Nuclear propulsion for space transportation has been on the drawing boards for decades. It will remain on the drawing boards and never be developed until NASA management eliminates its untenable launch vehicle operations costs. **We can continue to support a "jobs program" based on**

expendable, unaffordable, and unsafe launch vehicles that are going nowhere or we can support the CSS freighter and open the door to deep space.

Chapter 12 USAF Missed the Mark

USAF Space and Missile Systems Center Fumbles Again

"The Space and Missile Systems Center, a subordinate unit of USAF Space Command, is the center of technical excellence for developing, acquiring, fielding and sustaining military space systems. **SMC's mission is to deliver resilient and affordable space capabilities.** *The Launch Enterprise Directorate develops and acquires* **expendable launch systems** *and manages launch integration, mission assurance and launch campaigns."*

The above mission statement for the USAF Space and Missile Systems Center (SMC) identifies why this nation's space readiness to deter military and deep space threats is deficient. **Resilient and affordable space capabilities** cannot be attained with **expendable** launch systems. There is only one concept that can meet the SMC requirements and that's the CSS freighter. Regrettably, as the reader will learn later in this chapter, the SMC has rejected the freighter and is in process of continuing with their failed expendable launch system programs.

Paper Trail of Expendable USAF Launch Failures

The USAF SMC has had a tumultuous launch vehicle development history over the last three decades. In the early days of the Space Shuttle the USAF had started the development of their "Blue Space Shuttle" with planned launches from the Vandenburg west coast launch center, as shown below:

The shuttle was to be assembled in a building that covered the launch pad and just prior to launch the assembly building would separate for launch. This is the same concept proposed for the CSS freighter, which is one element for providing rapid launch capability. Unfortunately, the Challenger disaster and cost overruns caused by poor management led to the cancellation of the USAF's shuttle launch system plans. The USAF would abandon reusable launch systems and go back to their *expendable launch systems* policy. This has resulted in a long succession of failed expendable launch programs starting with the Titan IV launcher. The Titan IV launcher was developed from the family of the Titan intercontinental ballistic missiles.

Titan IV

Like the Space Shuttle, the launcher had two solid rocket boosters for launch. The Titan IV would launch 39 times with four failures. While the launch failure rate is not out of bounds for expendables, the launch vehicle average cost was nearly half a billion dollars and the payloads were sometimes $1 billion. Losing a billion dollar payload can always be expected when they are launched on expendable launchers...unlike the CSS freighter which has the capability to return the payload. The Titan IV launches would average less than 3 a year and the last launch was in 2005. Again it would be launch cost that would cause The Titan IV's demise.

Evolved Expendable Launch Vehicle Program

The next failed launch program was the Evolved Expendable Launch Vehicle (EELV) which was started in 1995. The goal of this program was to reduce launch cost by 25 to 50 percent by developing launch vehicles that would be competitive in the international market. Launching commercial payloads would have increased the number of flights which would have lowered launch cost. You've got to pay your operations staff the same annual salary if they launch one or ten launches per year.

The launch vehicles they finally selected were the Atlas 5, to be developed by Lockheed Martin, and the Delta IV, to be developed by the aerospace company now known as Boeing. Both launch systems would have varying configurations for launching different payload configurations and requirements.

The two companies failed to develop launch systems that could compete in the international market. At a space conference in Los Angeles, I asked a Boeing Delta IV executive why they failed to lower launch costs. Her reply was, "How can we compete when the foreign launch companies pay their workers $400 a month and we pay ours $400 a day?" I started to reply that we could compete with a reusable launch vehicle like the CSS freighter, but both the Boeing executive and I knew that the USAF management wouldn't listen. The Boeing executive also knew that their company could make a lot more money selling the USAF and NASA four or five expendable launchers per year rather than the same number of reusable freighters once every ten years.

To make the situation worse, the Lockheed Martin Company Atlas 5 developer would select the Russian's RD-180 rocket engine to power their first stage. The Russian RD-180 engine was and is far superior to any expendable engine that the U.S. has developed for its class. The approval for using the Russian engine was made with the understanding we would have the capability to produce a similar engine. By 2016 the U.S. rocket engine companies have not been able to duplicate the performance of the RD-180. As an engineering decision it was correct, but politically the decision was a disaster. More complications occurred when in 2006, the two competing companies decide to form a joint company called the United Launch Alliance. The USAF and NASA had to buy their launch services from a monopoly.

SpaceX Generates More USAF Problems

Space Exploration Technologies (SpaceX) has introduced a new innovation in the launch market, in that a private aerospace company has successfully developed a competitive commercial launch system. When SpaceX was refused the right to compete for a USAF launch contract because their Falcon 9 launcher was not "USAF certified", they sued. After a political firestorm, the Falcon 9 was certified in 2015. Another private contender called Blue Origin has entered the competition with a new rocket engine and launcher. United Launch Alliance was not to be left behind and announced their new launcher called the Vulcan. All these new launch systems have a common goal…some degree of reusability to "lower" launch cost. However, they are all still expendable launch systems and none can offer the unique capabilities of the CSS freighter, and eventually they will all fail. How many more failed launch programs can this nation afford?

USAF SMC Rejects the CSS Freighter

In November, 2014 I started another campaign to interest the USAF in the Space Shuttle reusable launch concept. My first letter was to Air Force Space Commander Gen. William L. Shelton to inform him of the progress that had been made with the CSS freighter and how it could solve the USAF's rapid launch turnaround needs, and significantly reduce their operation costs. I had a nodding acquaintance with Gen. Shelton when he was a junior officer assigned to the shuttle program in Houston. I don't believe my letter ever got to Gen. Shelton because I got an email reply from a USAF civil servant informing me that the USAF didn't have a need for a rapid turnaround launch capability, and they ignored the potential launch cost reduction. I suspect this was someone in their public relations office who screens letters to Gen. Shelton, and knew nothing about the USAF launch problems. I tried calling but couldn't get past the public relations barrier. One more example of how well intentioned but uninformed public relations civil servants are killing our space programs.

My next attempt was to Gen. Samuel Greaves, Commander Space and Missile Systems Center. The letter follows:

Nelson Aerospace Consulting
1407 Moller Rd
Alvin Texas
November 22, 2014

Lt. Gen. Samuel Greaves
Commander: Space and Missile Systems Center

As you are aware our space transportation for civil and military missions is in a dire state. I am writing on behalf of a small group of concerned aerospace engineers who believe that the only feasible and realistic solution is to develop a commercial space shuttle freighter to serve both the civil and military requirements. Our evaluations indicate that existing technologies can solve the problems of the decommissioned space shuttle and a commercially operated space shuttle freighter fleet could achieve the following:

- **Significantly reduce launch cost for** *the commercial, civil, and* **military customers.**
- **Provide stand-by on demand launch service in the event of military or deep space threats.**
- *Be the foundation for a space based transportation system for development of lunar and deep space resources.*

The NASA space shuttle was decommissioned because it was deemed to be too costly and unsafe. The commercial space shuttle (CSS) addresses these issues by:

- *Removing the civil service overhead, automating all flight phases (no pilots), encapsulated payloads, modular replaceable subsystems, and reduced contractor support.*
- *Installing crew escape pods on those missions transporting personnel to and from space. It is expected most CSS missions to be unmanned.*

Enclosed is an executive summary of the CSS. More details are provided at our webpage: www.spacetran21.org. The Aerospace Corporation and the America Institution of Aeronautics and Astronautics has been requested to review the technical feasibility of the CSS.

Gen. Greaves, we believe we are duty bound to provide space superiority for our nation. *As a former member of the NASA space shuttle team,* *you are aware of the potential of this launch system. China's Academy of Launch Vehicle Technology is also aware of the the space shuttle's potential (http://sinodefence.com/2014/02/22/chinas-spaceplane-programme/).*

I am requesting the opportunity to discuss the CSS with you and/or your staff.

Don A. Nelson
Aerospace Consultant (retired NASA engineer)
Coordinator Commercial Space Freighters

Somehow my letter got the attention of Gen. Greaves and he ordered his Advanced Concepts Division and Launch Enterprise Directorate to evaluate the CSS freighter. I wanted to have a face-to-face review process, but instead it was a number of telecom reviews.

The review process was off to a bad start and then it got worse. First, I had to convince them (mostly USAF civil service engineers who are the ones that really make the initial technical evaluations) that this was a concept evaluation and not a proposal for bidding for contract. Something I never accomplished because they kept telling me I had to fill out form such and such to be evaluated for contract. Then there were the diehard defenders of expendable launch vehicles who were convinced that the Space Shuttle launch system was unsafe and unaffordable. Then I got the regulations and rules folks who quoted gibberish like this:

*"In accordance with **MIL-STD-1540**, **Air Force space systems are subjected to** a combination of different verification methodologies (e.g. test, analysis, inspection) at several assembly levels (from units to the overall system) through the various program verification phases (e.g. Qualification, Acceptance, Pre-launch, On-orbit) to assure that launch and space equipment can function correctly and withstand stresses it may encounter during its life cycle. It is unrealistic to believe that any single construct or launch vehicle type can "eliminate" the potential of manufacturing defects or human operating errors that can impact mission success. Therefore, **SMC employs a comprehensive mission assurance strategy** that encompasses requirements, design, manufacturing, test, operations, personnel, safety, etc regardless of type of launch vehicle."*

I kept repeating it's a concept…judge it on the advantages it has over any expendable launch system. After these contentious telecoms I knew why SpaceX had to sue to get consideration for their Falcon 9 launch vehicle. You would think after three decades of failed expendable launch systems programs the USAF launch system engineers would be ready to try something new. But that's the problem. People resist change, and the CSS freighter launch system would certainly break some of the USAF civil service engineer's iron rice bowls (insured government jobs).

I got the following not unexpected "thank you…but" letter from Gen. Greaves in January 2016:

DEPARTMENT OF THE AIR FORCE
HEADQUARTERS SPACE AND MISSILE SYSTEMS CENTER (AFSPC)
LOS ANGELES AIR FORCE BASE, CALIFORNIA
27 January 2016
MEMORANDUM FOR NELSON AEROSPACE CONSULTING
FROM: SMC/CC
483 N. Aviation Blvd
El Segundo CA 90245-2808

SUBJECT: Commercial Space Shuttle Freighter Memo from Nelson Aerospace Consulting to Space and Missile Systems Center, dated 25 Nov 2015

1. Thank you for contacting the Space and Missile Systems Center (SMC) and for providing additional commentary and details on the subject concept for a Commercial Space Shuttle (CSS). At my request, the Advanced Concepts Division (SMC/AD) and the Launch Enterprise Directorate (SMC/LE) have reviewed your CSS Freighter concept. **Based on our review, SMC has determined that there are significant technical and logistical challenges with your concept and we believe that there are lower risk and more cost effective solutions for assured access to space. We will not pursue further the CSS concept study at this time.** *(See attachment USAF SMC review and my comments in my follow-up letter).*

2. SMC understands the evolving political climate, budgets, and launch service industry represent a major sea state change and with that we face significant acquisition and operational challenges. In spite of these challenges, we are committed to maintaining assured access to space by reducing launch costs, eliminating our dependence on the Russian RD-180 engine, and pursuing new entrant launch systems.

3. SMC encourages Nelson Aerospace Consulting to continue to explore options like your Commercial Shuttle and, in the event that you would like to submit a formal proposal, you may pursue open Federal Business Opportunity requests at https://www.fbo.gov/. You may also consider submitting your Commercial Shuttle or other related concepts in an Unsolicited Proposal (per FAR 15.6).

4. My point of contact for questions is Col XXXX XXXXX, available xxxx@us.af.mil or (310) 283-xxxx.
SAMUEL A. GREAVES
Lieutenant General, USAF Commander

Attachment: Bullet Background Paper on Commercial Space Shuttle Freighter Concept (see my reply comments to their conclusions in the following letter).

Nelson Aerospace Consulting
1407 Moller Rd., Alvin Texas
March 21, 2016
Lt. Gen. Samuel Greaves
Commander: Space and Missile Systems Center

Subject; Commercial Space Shuttle Freighter Review and Recommendation
Ref.: Commercial Space Shuttle Freighter Memo from SMC. Dated January 27, 2016

Gen. Greaves:

I would like to thank you for ordering the Space and Missile Systems Center (SMC) review of the Commercial Space Shuttle (CSS) freighter launch vehicle concept.

Unfortunately the SMC review team misinterpreted the CSS freighter concept as a proposed USAF project. The CSS can only be successful if developed in the private sector and specifically for the commercial market. **Our analyses indicate that the deficiencies of the NASA Space Shuttle can be corrected with existing technologies. The reusability, increased reliability, and payload recover capability are the key factors in making the CSS freighter a competitive launch system in the national and international launch markets.**

These are attributes the USAF space command's launch systems need but can never achieve with expendable launch vehicles. Every launch of any expendable launch vehicle is a fingers crossed situation that the flight will not experience a catastrophic failure caused by a "manufacturing error." The enclosed FAA chart shows that the overall reliability of expendable launchers has significantly improved over the decades but still experience a failure rate of five percent because of manufacturing errors. Only a reusable launch vehicle like the CSS freighter which has been flight tested can increase the reliability of launch systems.

Another serious concern with the SMC review team's evaluation is that the CSS freighter with its rapid launch turnaround capability makes it a resource to meet interstellar threats. However, the SMC review team has stated "countering interstellar threats is outside the charter of SMC launch services". This statement implies that USAF has no plans to come to the nation's defense in the event of an interstellar threat. Regardless, the USAF's counter measures would be extremely limited by the launch rate capability of expendable launch vehicles.

Like the SpaceX Falcon program, the CSS freighter uses existing technologies and with a proven airframe significantly reduces DDT&E cost. The SMC review team did not identify any unsolvable technology problems that would prevent the CSS freighter from being developed by a commercial enterprise. The USAF launch services would significantly benefit from a commercially operated CSS freighter. Private funding for the development would be accelerated if the USAF would publicly acknowledge their interest in commercial reusable launch systems. USAF current launch system developers and operators will not initiate a reusable launcher development until they realize the days of expendable launchers are over.

There must be a cultural change in the way the USAF procures launch services or the deficiencies of the obsolete expendable launch systems will continue to plague launch operations. I strongly urge you to endorse the commercial development of the CSS freighter and would welcome the opportunity to meet with you and discuss the advantages of the freighter.

Don A. Nelson
Coordinator Commercial Space Shuttle Freighter

The following answers SMC's reasons for rejecting the CSS freighter, an attachment sent in the Gen. Greaves' letter of January 27, 2016:

USAF SMC BULLET BACKGROUND PAPER ON COMMERCIAL SPACE SHUTTLE FREIGHTER CONCEPT PURPOSE
Reply Comments on USAF SMC Evaluation

The Nelson Aerospace Corporation has provided The Space and Missile Systems Center (SMC) with an overview and proposal for a Commercial Space Shuttle (CSS) freighter solution suitable for assured access to space and future National Security Space (NSS) launch services. SMC has examined the CSS freighter and identified several technical, logistical and fiscal challenges. This paper outlines several of these CSS freighter challenges.

DISCUSSION

- The CSS freighter design reference mission includes scope outside the charter of SMC launch services such as: human space flight, lunar missions, countering interstellar threats and recovering space debris.

Reply: *The CSS freighter is a commercial venture similar to that of the SpaceX enterprise. For USAF launch services to reduce their untenable launch cost they must adopt a launch system that supports commercial launch services which in turn reduces launch costs by increased launch rate.*

- Much of the CSS design concept is more suitable to NASA mission scope

Reply: *The CSS freighter's capability to support NASA's mission scope serves as another avenue to reduce launch overhead costs by increased flight rate.*

- SMC would require the CSS freighter to accomplish the new entrant certification process. Based on the antiquated technology in the existing shuttle system, the CSS freighter would require a significant amount of redesign and would therefore be a developmental program.

Reply: The SMC *"new entrant certification process"* is in itself antiquated. For example, a NASA study showed the cost to develop the SpaceX Falcon 9 using the NASA-USAF Cost Model *(NAFCOM)* would have increased the development costs by a factor of three. The USAF can expect the proposed Vulcan to suffer from the same excessive development cost problems and still be an obsolete vehicle with inherent manufacturing problems.

- Based on significant rework, the Nelson Aerospace proposed cost of $200M per launch does not account for the non-recurring costs associated with a new development program.

Reply: The CSS freighter launch concept is based on a *"ship and shoot'"* business model. This transfers a significant portion of the non-recurring development cost usually assigned to the launch system to the payload designer (see following for more explanation).

- Based on NASA's original Shuttle Program cost of $200B, CSS redevelopment could cost $50-$200B

Reply: This conclusion would be valid if the CSS freighter was developed using the NAFCOM program approach, WE CANNOT AFFORD TO CONTINUE ON THIS PATH!

- This cost does not account for new launch, landing, and operations locations, facilities, equipment, staffing and infrastructure

Preliminary estimates indicate that the infrastructure and vehicle development cost are in the $10b to $12b range. Staffing is in the low one hundreds range. This is in the support cost range for a privately developed venture.

- This cost does not account for a new upper stage which would also be required to transfer satellites into higher orbits beyond Low Earth Orbit

Reply: The CSS freighter is the first step in establishing a completely reusable space transportation system. Until the space based low orbit vehicle can be commercially developed existing upper stage can be used and possibility recovered for reuse.

- SMC assesses that developmental cost for the CSS freighter would be orders of magnitude greater than recent SMC estimates for a next generation EELV launch system.

Reply: Again the SMC cost models do not account for the saving of a launch system developed in and for the commercial market.

- *The CSS proposal is based on the reuse of Space Shuttle Main Engines*
- *Existing Shuttle Main Engines are dedicated to the NASA Space Launch System*

Reply: *The Space Shuttle RS-25 is an existing engine that can still be manufactured. There are existing significant improvements to that engine which can be incorporated in the new engines.*

- *The EELV program is a launch services procurement and, while we support reusable launch system solutions, reusable technologies must be driven by and funded by the launch service provider*

Reply: *The EELV program(s) has failed and will continue to fail to reduce launch costs because these vehicles cannot compete in the international launch market. The EELV's requires multi-levels of expensive backup systems to provide reliability, yet they still suffer from manufacturing error.*

- *Aerospace analyses indicates that the most cost effective reusable launch system incorporates a reusable first stage versus a reusable second stage*

Reply: *Many in the aerospace community have concluded that it is not cost effective to recover the first stage of a launch system derived from an expendable launch vehicle. Though not cost effective the NASA Space Shuttle was a proven reusable launch vehicle.*

- *Commercial industry partners, such as Blue Origin and SpaceX, have also determined that a reusable first stage is a very effective approach to reducing launch costs and the Air Force enthusiastically supports commercial efforts to develop the enabling technologies*

Reply: *Studies conducted by NASA in the 1990's timeframe concluded that recovery of first stage components of expendable launch vehicles were not cost effective. There have been no technical advances to make this conclusion invalid.*

SUMMARY

SMC has developed a comprehensive strategy for maintaining assured access to space. SMC is executing operations, mission assurance plans, and acquisitions to reduce launch costs, eliminate dependence on the Russian RD-180 engine, and pursue new entrant launch systems. SMC has reviewed the CSS proposal and determined that there are many technical, logistical and fiscal issues that make the CSS solution higher risk than the planned services and future concepts in the current investment strategy.

Reply: SMC comprehensive strategy for maintaining assured access to space is based on an obsolete concept. To face the challenges of acute budget restrictions, launch operations costs must be significantly reduced. Payloads costing over $1b cannot be flown on launch systems that have inherit manufacturing errors and cannot be recovered in the event of a failure. To continue with the strategy of expendable launch vehicles only insures failure.

There was no reply to this letter. The USAF is now on a course to repeat decades of failed expendable launch vehicle programs. I believe history will show that Gen. Greaves and the U.S. taxpayer got "had" by a group of civil service engineers who failed to conduct an unbiased evaluation of the CSS freighter.

In my opinion continuing to develop any USAF expendable launch system is a failure of due diligence, and I reported this breach of responsibility to the Department of Defense inspector general office in the following letter:

Concerned American Aerospace Engineers

Don a. Nelson, Coordinator

1407 Moller Rd Alvin Texas 77511-3248

May 13, 2016

Mr. Glenn A. Fine

Acting Inspector General

U.S. Department of Defense

Office of Inspector General

4800 Mark Center Drive

Alexandria, VA 22350-1500

Subject: USAF Space and Missile Systems Center Due Diligence Investigation Request

Dear Mr. Fine:

The USAF's ability to maintain assured access to space and reduce launch cost cannot be achieved by their plan to continue with expendable launch vehicles like the proposed United Launch Alliance (ULA) Vulcan launcher. Every launch of an expendable launcher is a "flight test" of a vehicle that is subject to manufacturing errors. A Federal Aviation Agency (FAA) study (enclosed) of expendable launchers found that a 5 percent annual failure rate is not uncommon. Proposed plans to recover and reuse first stage engines and components of expendables to reduce launch cost remains unproven and there is no justification to believe that USAF launch plans will significantly reduce launch costs. To continue with reliance on expendable launch vehicles is a path of continuing with unacceptable launch costs and uncertain reliability.

There is only one avenue for reducing launch costs and improving reliability and that is a reusable launch vehicle. The NASA Space Shuttle failed to achieve these objectives primarily because of mismanagement. The Space Shuttle design did not consider the need for eventual subsystem upgrades and as a result a simple flight subsystem change could take months and cost millions of dollars. When a presidential directive prohibited commercial payloads being launched on the shuttle, the flight rate dropped and the launch cost soared. With the loss of two flight crews the Space shuttle was declared not only unaffordable but unsafe for human transportation and was decommissioned. NASA management avowed that the shuttle replacement would be 10 times safer; but when challenged, the NASA Aerospace Safety Advisory Panel had to admit that the Space Launch System (having expendable stages with inherent manufacturing errors) "was not significantly safer than the Space Shuttle."

Since the decommissioning of the NASA Space Shuttle, a small group of concerned American aerospace engineers have worked to resolve the shuttle's technical and operational problems. Existing technologies have been identified which solve the NASA shuttle's problems and significantly improve operations. Using these innovations, a plan has been formulated for developing a commercial Space Shuttle (CSS) freighter which will provide more cost effective and reliable launch services for commercial, civil, and military access to space. Basically the CSS freighter reduces launch cost by eliminating the costly manufacturing of expendable launchers and increases flight rate by offering a unique five day launch turn-around capability. Payload insurance rates are significantly reduced because of the increased reliability in reducing manufacturing errors and the CSS freighter's capability to recovery the payload.

The CSS freighter's attributes are essential to the USAF's requirement for affordable assured access to space. In November 2014, Lt. General Samuel Greaves, Commander Space and Missile Systems Center was informed of the CSS freighter plan. Gen. Greaves recognized the potential of the CSS freighter and ordered his staff to conduct an evaluation. Unfortunately the CSS freighter plan met the same cultural resistance from the review team as was the case for the contentious SpaceX launch vehicle proposal. In a January 2016 memorandum (enclosed) they replied: "Based on our review, SMC has determined that there are significant technical and logistical challenges with your concept and we believe that there are lower risk and more cost effective solutions for assured access to space." **In other words SMC wants to continue with the failed Evolved Expendable Launch Vehicle program concept. Ironically their attached "Bullet Background Paper" review (enclosed with comments) did not identify any "significant technical and logistical challenges."**

The CSS freighter can only be successfully developed by the private sector (with private funding) and specifically for the commercial market. The only commitment from the USAF would be a statement that if the CSS freighter achieved its design goals the USAF would be a buyer of launch services. This commitment would be significant in obtaining private funding for the CSS freighter development. Furthermore it would put on notice to the aerospace organizations profiting from manufacturing expendable launchers that the era of these obsolete launch vehicles is ending.

Will the Office of the Inspector General be silent on this grievous failure of responsibility?

Don A. Nelson

Coordinator, Concerned American Aerospace Engineers

CC:

Secretary of the USAF, Deborah Lee James

Under Secretary of Defense for Acquisition, Mr. Frank Kendall

USAF Space Commander, Gen. John E. Hyten

U.S. Government Accountability Office

There was no reply to this letter. In all likelihood, this letter was never read by the Inspector General. **The USAF** *"resilient and affordable space capability"* **continues on a path of more certain failures.**

Note: In 2011 NASA conducted a development cost comparison study for the privately developed SpaceX Falcon 9 launch vehicle. They used the NASA-Air Force Cost Model (NAFCOM). NAFCOM is the primary cost estimating tool that NASA and the USAF uses to predict the costs for launch vehicles. The cost model predicted the Falcon 9 launch vehicle would have cost between $1.7 billion and $4.0 billion if developed by NASA or the USAF. SpaceX developed the Falcon 9 for $390 million. NASA has reportedly verified the SpaceX development costs.

The CSS freighter privately developed launch vehicle would cost a fraction of what the USAF is planning to spend on their next "obsolete" expendable launch vehicle.

Chapter 13 Today

"I think it's fair to say that there's been a sense of drift to our space program over the last several years"…President Barack Obama

Today there is still a *drift to our space program* but the *drift* has been for several decades and not just *the last several years*. I attribute this *drift* to a failure of due diligence by NASA's human space flight management and the NASA oversight panels and agencies charged with the responsibility of investigating and reporting problems. If I am correct, then why has the *drift* or failure of due diligence been allowed to continue for decades? I believe it was not just the NASA management and oversight panels that allowed our human space endeavors to drift…there are other groups and individuals that have failed in their responsibility.

First let me reiterate the list of significant human space transportation failures:

- Space Shuttle Decommissioned –Without commercial payloads the flight rate never increased and launch cost soared. Having both contractor and civil service personnel operating the shuttle launch system created a government "job program" where the shuttle operations work force grew to over 21,000 employees. I believe this bloated work force was an indirect cause of both Space Shuttle disasters, because the launch operations were too large to manage. In my professional opinion, the commercial Space Shuttle workforce could eventually have been reduced to 3,500 or less employees.

- X-33 VentureStar Canceled – Right concept, wrong vehicle. The VentureStar was to be a commercial launch system endeavor developed by a joint venture with NASA and an aerospace company. The program had the possibility of succeeding if NASA management had not imposed the requirement of a single stage launch to orbit which required developing unproven and unachievable technology. It was canceled in 2001 due to technical problems and failure to control the vehicles weight growth.

- Constellation Canceled – Promoted as a sustainable, affordable, and safe human space transportation… failed because it was based on expendable launch vehicles, which can never achieve any of those goals. Canceled in 2009.

- Space Launch System – One more time a funded sustainable, affordable, and safe Space Shuttle replacement which isn't. This program was initiated as a ploy to avoid the political fallout in the cancellation of the Constellation program. It resulted from NASA management's failure to concede that heavy-lift expendable launchers and crew modules are not affordable and are not safe. Oversight panels have remained silent. Cancellation is TBD, the sooner the better for the U.S. taxpayer.

So why are NASA's human space flight managers being allowed to continue down a path of certain failure and cancellation? Those in the aerospace community and many in the general public get excited about the prospect of returning to the Moon and taking humans to Mars. NASA's public relations program office does a good job promoting human space exploration. What is not being told is that we can't afford to go back to the Moon, and especially to Mars, with expendable space transportation vehicles. The technology we have today enables the development of reusable space shuttle freighters and reusable space based cruisers for near earth and deep space operations. So why is NASA human space flight management still today being allowed to cling to their obsolete and unsafe 20th century expendable space transportations systems? Could it be no one or no group wants to be responsible for pulling the plug on NASA's human space program, or is it they don't understand the problem? I think it's a little of both.

Crew Safety Today Still Unsafe

With the loss of two shuttle crews, NASA management has "played the safer card" in their attempts to promote the Orion and commercial crew modules. At first they told us the potential loss of a crew was only 1 in 2000 missions for the crew modules. Today the safety prediction number is a

goal of 1 in 270. By now the reader is aware I've challenged NASA's crew safety predictions for years. The numbers are computed by a risk analyses program that fails to account for failure during the entry phase where crew modules have no crew escape systems. It is a crew escape system that saves a crew, not risk analyses computer programs. **It is my opinion that these bogus safety predictions are another major contributor to our human space programs continuing failures.**

Another NASA Safety Panel Challenge

At a NASA safety panel meeting in January, 2014 I presented following information which showed NASA space flight management has misrepresented the safety of the SLS/Orion crew module. The following are excerpts from the presentation:

NASA Release 11-164:
Headquarters, Washington
202-358-1979/5241
michael.j.braukus@nasa.gov j.d.harrington@nasa.gov
May 24, 2011
RELEASE : 11-164
NASA Announces Key Decision For Next Deep Space Transportation System
WASHINGTON -- NASA has reached an important milestone for the next U.S. transportation system that will carry humans into deep space. NASA Administrator Charles Bolden announced today that the system will be based on designs originally planned for the Orion Crew Exploration Vehicle. Those plans now will be used to develop a new spacecraft known as the Multi-Purpose Crew Vehicle (MPCV).
"We are committed to human exploration beyond low-Earth orbit and look forward to developing the next generation of systems to take us there," Bolden said. "The NASA Authorization Act lays out a clear path forward for us by handing off transportation to the International Space Station to our private sector partners, so we can focus on deep space exploration. As we aggressively continue our work on a heavy lift launch vehicle, we are moving forward with an existing contract to keep development of our new crew vehicle on track."

*Lockheed Martin Corp. will continue working to develop the MPCV. The spacecraft will carry four astronauts for 21-day missions and be able to land in the Pacific Ocean off the California coast. The spacecraft will have a pressurized volume of 690 cubic feet, with 316 cubic feet of habitable space. It **is** designed to be 10 times safer during ascent and entry than its predecessor, the space shuttle.*

NASA release 11-164 states the Orion Multi-Purpose Crew Vehicle, "is designed to be 10 times safer during ascent and **entry** than its predecessor, the space shuttle". **In actuality, the Orion crew module and all the commercial crew modules are potential death traps because they have no escape systems (pods) for the re-entry phase of flight.** I challenged the NASA Office of Safety and Mission Assurance on the authenticity of this release. Their reply indicated they were unaware of this release and confirmed that the Orion crew module had failed to meet the Constellation preliminary design requirements for entry.

From: Don Nelson [mailto:danelson@wt.net]
Sent: Wednesday, June 15, 2011 9:04 PM
To: XXXXX XXXXXX
Subject: Crew Module Safety Issue
To:
Office of Safety and Mission Assurance
Aerospace Safety Advisory Panel
Subject: Crew Module Safety Issue
*Recently published statements attributed to NASA state that the Multi-Purpose Crew Vehicle Is: "designed to be 10 times safer during ascent and **entry** than its predecessor, the Space Shuttle." Is this statement the position of the NASA's safety oversight authorities? If so, as a retired NASA engineer with extensive experience in the operation of crew modules, I challenge the authenticity of this statement.*
While the crew escape tower on the MPCV does provide significant improvement over a Space Shuttle without crew escape pods, it does not negate the many factors that have made crew modules a death trap during the re-entry phase of flight. As example, historically the Russian Soyuz crew module's safety record is not significantly better than that the Space Shuttle. While the Soyuz crew module has experienced a failure of

the escape tower, it has been the re-entry phase of flight that has proven to be the fatal environment for flight crews. Potential fatal crew module failures are:

- *Every crew module flight is a test flight! Manufacturing errors have occurred.*
  http://www.aviationweek.com/aw/generic/story.jsp?id=news/Russia 052208.xml&headline=Soyuz%20Ballistic%20Re-entry%20Explained&channel=space)

- *Water landings are an unacceptable dangerous risk to flight and recovery crews. Land landings have the potential to expose the crew to fatal high g loads upon impact.*

- *Crew modules have very limited cross range capability which could require a reentry into unacceptable weather conditions.*

- *Crew module's notorious reentry errors result in an expanses landing zone that could prevent rapid access to the crew in dire circumstances.*

- *Parachutes are known to fail. This is another unacceptable single point failure.*

There are too many potential failures with fatal consequences for a crew module to be even considered for 21ˢᵗ century human space transportation. The Russian Soyuz crew module is still in service only because their government cannot afford to develop a safer reusable lifting body winged runway landing crewed spacecraft (http://en.wordpress.com/tag/kilper/). **While NASA spends billions developing the MPCV crew module which is nothing more than a political derived government jobs program. Furthermore, the crewed missions proposed for the MPCV can be conducted more efficiently with robotic spacecraft. The silence of the NASA crew safety oversight authorities on the safety of the MPCV is a deadly silence. Will you remain silent?**

Don A. Nelson
Retired NASA Aerospace Engineer

Today

NASA Safety Office Reply:

From: *XXXXX XXXXXX*
To: *"Don Nelson" <danelson@wt.net>; "Burch, Susan (HQ-TD000)" <susan.burch@nasa.gov>*
Cc: *"Cooke, Douglas (HQ-BA000)" <douglas.cooke-1 @nasa.gov>; "Wilcutt, Terrence W. (JSC-NA111)" <terrencexxx xx@nasa.gov>*
Sent: Thursday, June 16, 2011 6:40 AM
Don,

I don't know about these statements (or misstatements). I cannot vouch for everything NASA people have said over the past few years about risk, but there is nothing official out of NASA saying Orion or MPCV must be ten time safer than shuttle for ascent or entry. The level 1 requirements set for Cx included a requirement that the PRA show total ascent risk to be 1/1000 in the mean, and same for entry. Those numbers represent somewhere between half and 1/3 the risk of shuttle for the same phases of flight.

At PDR, the Cx design PRA estimate was better than the requirement for ascent **and not there yet for entry***, but they had some design changes, including improvements in landing system failure tolerance they were looking at to get the entry PRA to 1/1000.*

We don't have a set of level 1 requirements yet for the next NASA developed human system, but we do plan to use the Cx numbers above as part of our human rating requirements set for commercial crew to ISS. **I agree this will be a challenge for any capsule for all the reasons you give if not more.**

As for my community's silence on this matter, I normally don't comment on unsubstantiated rumors floating around the internet, but be assured I have a voice and an audience within NASA, and I will not hold my tongue if I think we are doing something that puts our crews in an unacceptable risk posture.

Best,
XXXXX

Any competent NASA human space flight manager or engineer should have known that crew modules have unsolvable entry problems and are no safer than the Space Shuttle. **They had the moral responsibility to speak**

208

out...they all remained silent! Instead of a safer CSS freighter "today" we continue to spend billions of dollars developing unsafe Orion and commercial crew modules for human space transportation. Is there any question why our human space programs continue to fail?

Did NASA Safety Panel Hide Their Finding?

Why didn't the following cover letter for the NASA safety panel's annual report alert the Congress to their findings that the SLS/Orion has an unsolvable safety issue, and is not significantly safer than the decommissioned Space Shuttle?

NASA AEROSPACE SAFETY ADVISORY PANEL
National Aeronautics and Space Administration
Washington, DC 20546
XXXXX, Chair
January 28, 2015
The Honorable Joseph R. Biden
President of the Senate
Washington, DC 20510
Dear Mr. President:
Pursuant to Section 106(b) of the National Aeronautics and Space Administration Authorization Act of 2005 (P.L. 109-155), the Aerospace Safety Advisory Panel (ASAP) is pleased to submit the ASAP Annual Report for 2014 to the U.S. Congress and to the Administrator of the National Aeronautics and Space Administration (NASA).
This Report is based on the Panel's 2014 fact-finding and quarterly public meetings; "insight" visits and meetings; direct observations of NASA operations and decision-making; discussions with NASA management, employees, and contractors; and the Panel members' past experiences.
The ASAP applauds NASA's accomplishments during this past year. These include safe International Space Station (ISS) operations, growing traction on the Exploration Systems Development programs, success in supporting ISS logistics via commercial cargo, and positive strides in infrastructure management.

Regrettably, the Panel is unable to offer any informed opinion regarding the adequacy of the certification process or the sufficiency of safety in the Commercial Crew Program (CCP) due to constraints on access to needed information.

In this Report, we note that NASA is experienced and accomplished in procuring space systems by "making" (e.g., NASA custom-produced satellites), "managing" (e.g., a NASA program office managing fulfillment of a "performance spec," often designed and generally produced by a contractor), and "buying" (where the marketplace has established the bona fides of value, safety, and reliability).

The CCP falls within a chasm between the deep insight of "managing" and that of "buying" a product proven by broad market acceptance. With the CCP, NASA is operating at relative arm's length while concurrently fostering the development of a commercial market. The Panel strongly believes open communication and transparency are essential to ensuring the safety of the program.

NASA's senior leaders and staff members offered significant cooperation to support the completion of this document.

I submit the ASAP Annual Report for 2014 with respect and appreciation.

Sincerely,

XXXXX, Chair, Aerospace Safety Advisory

The NASA safety panel in this and later reports failed to alert the Congress that NASA was and still is mispresenting the crew safety of the SLS/Orion flight system. Instead, they buried their admission in page 7 of this report with this one sentence excerpt **"SLS Orion programs...are not significantly safer than the actual historical performance of the Space Shuttle"** in the body of the report where members of Congress or their staff never saw it or failed to understand its importance. Also, the NASA safety panel tried to blame the Public Affairs Office for NASA's management deception. See the following, from report:

*"Because Probabilistic Risk Assessment results provide a risk assessment of the design capability at maturity, actual risks for early operations of the Space Launch System (SLS) and Orion could be significantly higher than the calculated or "advertised" risk. Because the perception of external stakeholders is vitally important, **NASA's Office***

of **Communications** *must be cautious not to create or reinforce inaccurate perceptions of risk."*

The NASA safety panel should have known that the Office of Communications did not make this false safety assertion. That office did not conduct the risk analyses; that information was from NASA's human space flight management.

In the following 2015 NASA safety panel report, they continue to misrepresent the safety calculations:

"The follow on Constellation (that was originally envisioned to replace Shuttle) had a goal (Loss of Crew…LOC) of 10 times better (1 in 1000), based on a 2005 study, which at that time was thought possible and was consistent with the request from the Astronaut Office. As the Constellation system design began and the program started looking at hazards and threats, in particular the very significant Micrometeoroid and Orbital Debris (MMOD) threat, **they found that 1 in 1000 was going to be an impossible number to meet. The Agency decided toward the end of that program to reduce the Loss of Crew (LOC) number to 1 in 270 (or 3 times better than Shuttle at end-of-life)***. The Panel had a concern with that LOC number at prior meetings and asked NASA to relook at it. They did and felt it was the best that could be done. When the CCP came along, the HEOMD chose that same number to keep an even playing field (commercial crew should be as safe as* **Constellation would have been), and the requirement of 1 in 270 was set for commercial crew.**

The above NASA safety panel minutes verify that the panel still endorses a crew safety level for the commercial crew modules that cannot be achieved because it has the same safety flaws as the SLS/Orion crew module…manufacturing errors and no re-entry crew escape system. In the NASA safety panel's 2016 annual report, loss of crew risk was still not identified as not significantly better than the Space Shuttle. The NASA safety panel should read their own 2014 annual report where they state, *"the Loss of Crew probability thresholds for them (SLS/Orion) are not significantly safer than the actual historical performance of the Space Shuttle."*

The reader is reminded that **today's** NASA safety panel still does not conduct any analyses and only evaluates what NASA human space flight management tells them! Until the NASA safety panel is made into an independent panel and not subjected to NASA management approval, there will be no unbiased safety evaluations.

Should the NASA Safety Panel Members Resign?

I wrote the following letter to the panel members which questioned their capability to perform their duties.

Concerned American Aerospace Engineers
Coordinator: Don A. Nelson
Alvin Texas   June 6, 2017

NASA AEROSPACE SAFETY ADVISORY PANEL
Washington, DC
Dr. Patricia Sanders, Chair and Panel members

Subject: NASA Safety Panel Resignation
The ineffectiveness of the NASA Aerospace Safety Advisory Panel has resulted in our astronauts being transported in unsafe and obsolete crew module spacecraft. However, what is more concerning is that this failure to provide the astronauts with the safest obtainable space transportation has denied the earth's population protection from the increasing asteroid/comet impact threat.

The Panel was established by Congress in 1968 as a result of the January 1967 Apollo 204 crew command module fire that killed three astronauts. Their principal statutory duties are to review NASA's programs and operations plans and advise NASA and the Congress as to the hazards of proposed plans and current operations. The intent was to have an independent safety review; however this objective has never been achieved and was a contributing factor in the loss of the Challenge and Columbia Space Shuttle crews and now has our astronauts and all of us on spaceship Earth in a dangerous position.

It is the lack of being a truly independent safety evaluator that has resulted in the Panel's ineffectiveness and failures. The Panel is appointed by and serves at the pleasure of the NASA Administrator. The Panel does not conduct its own investigations but relies on NASA to inform it of how the plans and procedures for safety operations are evaluated. In general, when the Panel is presented with safety concerns from an outside source they are ignored. Without independent verification capability the Panel's decisions are based only on their own conjecture and have no supporting information to challenge NASA's position.

For example, the Space Shuttle was allowed to be decommissioned because NASA human space flight management conducted loss of crew analyses that predicted a crew module spacecraft would be ten times safer. When the Panel was given irrefutable information that the NASA crew module safety evaluations were in error they issued a statement in their annual report to Congress that NASA's crew modules were not significantly safer than the Space Shuttle. However, the revelation that NASA had misrepresented the safety of their crew modules spacecraft was blamed on the NASA public affairs office and the disclosure was buried in the interior of the report. Later *Panel reports continued to support NASA's untrue position that their crew modules are safer than the decommissioned Space Shuttle.* The Panel has also been made aware of significant technology advancements that can be incorporated in an advanced space shuttle concept that permit the installation of crew escape pods which provides significant improvement in crew safety. However, the Panel has continued to remain silent.

The development of a safer and more affordable advance space shuttle concept has been thwarted by the Panel's continued endorsement of crew modules. The development of the advance space shuttle has become imperative as we learn of the increasing possibilities of an asteroid/comet impact on Earth. The advanced space shuttle concept (also known as the Commercial Space Shuttle freighter) is the only launch system that can provide the rapid and numerous launches required to counter a possible impact threat.

The unsafe and obsolete crew modules will continue to transport our astronauts and spaceship Earth's passengers will continue to be exposed to the danger of an impacting asteroid/comet until there is an event that exposes the failures of the NASA safety Panel. If the

members of the Panel are truly dedicated, their best avenue for correcting this path of certain disaster is to resign!
Don A. Nelson
Coordinator, Concerned American Aerospace Engineers

No reply.

Office of Special Counsel Redefines Gross Mismanagement

The Office of Special Counsel has responsibility for investigations of government gross mismanagement. Their charter reads as follows:

"The United States Office of Special Counsel is a permanent independent federal investigative and prosecutorial agency whose basic legislative authority comes from four federal statutes, the Civil Service Reform Act, the Whistleblower Protection Act, the Hatch Act, and the Uniformed Services Employment and Reemployment Rights Act. The agency also operates a secure channel for federal whistleblower disclosures of violations of law, rule, or regulation; gross mismanagement; gross waste of funds; abuse of authority; and substantial and specific danger to public health and safety."

You have to be an active or retired government employee to request an investigation from this office. With the NASA's human space program in shambles, I requested Office of Special Council to consider if NASA space flight management had mismanaged the Space Shuttle decommissioning and the subsequent failed human space programs. My first letter to the office follows:

April 13, 2012
The Honorable Carolyn Lerner
Head, Special Counsel
1730 M Street, N.W., Suite 218
Washington, D.C. 20036-4505
Subject: NASA and NASA Oversight Offices Failure of Due Diligence
Ms. Lerner:

The decommissioning of the space shuttle launch system was conducted under the premise that it was an unaffordable and unsafe launch system. Enclosed in this letter is documentation that NASA management has misled the President and Congress about the capability and safety of the launch systems that they were/are developing to replace the space shuttle.

I won't publish the total letter or the attachments because they reiterate what has already been presented in this book. My question was, did NASA management make every **reasonable** effort to insure that the Space Shuttle launch system replacement was a more affordable and safer human space transportation system? NASA management has wasted roughly some $70 billion in failed and failing efforts to revive the human space program with expendable launch vehicles and crew modules. Was this caused by gross mismanagement or gross incompetence? These were was the questions I asked the Special Counsel to investigate. Their reply follows:

U.S. OFFICE OF SPECIAL COUNSEL
1730 M Street, N.W., Suite 218
WASHINGTON, D.C. 20036-4505

October 16, 2012

Mr. Don A. Nelson 1407 Moller Road Alvin, TX 77511

Re: OSC File No. PI-xxxxxxx
Request for Reconsideration

Dear Mr. Nelson:

The Office of Special Counsel (OSC) has completed its review of the information you referred to the Disclosure Unit. You alleged gross mismanagement and a substantial and specific danger to public safety by employees at the National Aeronautics and Space Administration (NASA), Washington, District of Columbia.

OSC is authorized by law to refer protected disclosures to the involved agency for an investigation and report. Disclosures that OSC may refer for investigation must include information that establishes a substantial likelihood of a violation of law, rule, or regulation, gross mismanagement, a gross waste of funds, an abuse of authority, or a substantial and specific danger to public health or safety... **OSC does not have the authority to investigate disclosures and, therefore, does not conduct its own investigations.**

You alleged that management officials failed to exercise due diligence by choosing to utilize the Heavy Lift Space Launch System (SLS) for the Human Deep Space Program without adequately considering the alternative option of privatizing the space shuttle fleet. In addition, you contended that the agency's failure to provide information responsive to a Freedom of Information Act (FOIA) request you submitted on September 9, 2010, "requesting the findings of NASA studies that verify that their current plan to develop a heavy lift launch system and/or purchase commercial launch services ... will be safer and more cost effective than a spaceflight transportation system ..." indicates that management officials "never conducted any studies to verify that the launch vehicle(s) they chose would be as cost efficient or safer than the existing or privatized space shuttle." While you believe that privatization is safer for shuttle crew and more cost effective, you stated in your disclosure that you did not wish for OSC to consider whether privatization is superior to the SLS. You specified that you only intended for OSC to review your allegation regarding NASA's failure to adequately compare the two options.

Upon review and consideration of the information and documentation you provided, it appears that the issue you have raised concerns a matter of agency policy and discretion. Although you strongly object to management officials' decision to utilize the SLS in lieu of privatizing the space shuttle fleet, we are unable to find that the agency has acted outside of its discretion in doing so. Furthermore, while you believe that privatization is the safer and more cost effective option, differences in opinion are not sufficient to establish a substantial likelihood of wrongdoing.

In addition, we cannot conclude that the agency's lack of information responsive to your FOIA request established that it did not evaluate the relative merits of the SLS and privatization. On the contrary, it appears that NASA has considered privatization based on an April 1, 2010, letter from XXXX XXXXXX, Associate Director for Space Operations, which you provided to our office. In that letter, Mr. XXXXXX explained that privatization is "highly dependent on market demand" and that "there may not be the market demand to profitably sustain a commercially-operated Space Shuttle." Thus, it seems that NASA did evaluate your concerns. While you may contend that a more formalized review process is necessary, you have not identified, nor are we aware of, a law, rule, or regulation that requires NASA to follow a formal procedure of comparing available options prior to making decisions of this nature.

Consequently, for the above reasons, **we cannot determine with a substantial likelihood that gross mismanagement, or a substantial and specific danger to public safety has occurred.** *As a result, we will take no further action with respect to this allegation.*

Should you wish to pursue this matter further, you may contact the National Aeronautics and Space Administration, Office of Inspector General, 300 E Street, S.W., Code W, Room 8V39, Washington, D.C. 20546; main telephone number: (202) 358-1220; hotline number: (800) 424-9183.

Accordingly, we are closing our file. If you wish to discuss this matter, please contact me at (202) 254-XXXX.

Sincerely,
XXXX XXXXXX
Attorney, Disclosure Unit

The reader will note in the OSC's conclusion that "*we cannot determine with a substantial likelihood that gross mismanagement or a substantial and specific danger to public safety has occurred*" is based on statements containing *it appears*, *may not* and *it seems*. It is ironic that the OSC letter referred me back to the NASA Office of Inspector General, which had already confirmed that they had *no records* that NASA management *conducted any studies to verify that the launch vehicle(s) they chose would be as cost efficient or safer than the existing or privatized space shuttle.* The reader is reminded that the OSC *does not have the authority to investigate disclosures and, therefore, does not conduct its own investigations.* **Once again, biased conjecture triumphs when no investigative analyses are conducted.**
Like other oversight agencies, I believe the OSC decision was based on the belief that NASA management would always be responsible and do its homework. They don't do their homework, as the long list of failured human space programs confirms.

Over the years I have continued to challenge the OSC to reconsider NASA's gross mismanagement. Today, in 2017, the OSC still holds to their position.... *that incompetence does not equate to gross mismanagement.*

Note: Again the reader is asked to judge that if President Bush had been informed that any shuttle replacement expendable launch vehicle with crew modules would not be significantly safer than the Space Shuttle, would he have ordered it decommissioned? In making that decision, the reader is reminded that many of the shuttle's missions could have been flown unmanned. **Did the OSC err in their decision?**

Note: In August 2017, the OSC was sent a manuscript copy of this book and asked again to evaluate if NASA's space flight management misrepresentation of crew safety has resulted in the waste of billions of dollars and exposed our astronauts and all of us on space ship Earth to unwarranted risk.

Civil Service Morass

History has shown us that the demise of great empires began with the failure of their civil service systems. Today, it is a common occurrence to learn of failures in our civil service system. Failures range from letting us drink water contaminated with lead to remaining silent as unethical bankers put this nation on the threshold of bankruptcy. This failure of responsibility does not go unnoticed by the young NASA civil service scientists and engineers who we expect to dedicate their careers to securing mankind's future in space. I don't see the same enthusiasm for *today's* human space endeavors as I saw in the Apollo and Space Shuttle programs. When I worked these programs, the parking lots at the Johnson Center were always full on Saturdays and nearly half-full on Sundays. I worked many sixty hour weeks and a lot of that time was voluntary with no pay compensation. I never heard any complaining because what we were doing was exciting and meaningful work. Today, the Johnson Center parking lots are virtually empty on weekends and flex Fridays. Just having flex Fridays is an indicator that the work is not exciting and meaningful.

Today the Aerospace Industry Still Remains Silent

Too often I have been approached by contractor aerospace engineers who have expressed their concerns for the safety and excessive cost of the Orion and the expendable launchers. However, they remain silent because to bring up problems that cannot be solved would result in program cancellation and could cost them their jobs. Executive industry management remains silent because the profit from continued manufacturing of expendable launch vehicles far exceeds that of a reusable launch system. This is a deadly silence. Today, this nation cannot afford to support the manufacturing of expendable space vehicles and develop the space vehicles needed for deep space transportation. While our aerospace launch industry remains complacent, the competitive foreign launch industry has already developed launch systems whose capabilities meet or exceed that of this nation. The launch of the Chinese Long March V in November 2016 matches the payload lift capability of our Delta 4 heavy launcher.

Today's NASA Oversight Panels Still Remain Silent

In a continuing effort to get NASA's oversight panels to acknowledge that NASA's expendable human launch systems can never be sustainable, affordable, and safe, I presented the following to the NASA Advisory Council in December 2015:

NASA Advisory Council
Public Input 12/3/2015
Don A. Nelson caae@wt.net
Retired NASA Engineer – Coordinator Commercial Space Shuttle Freighter

I represent a group of aerospace engineers who believe the SLS/Orion will be canceled. NASA should prepare for that eventuality and encourage the development of a Commercial Space Shuttle freighter. www.spacetran21.org

Why SLS/Orion will be canceled:

Expendable Launch Vehicle Failure Rate Unacceptable
(Presented to GAO August 2015)
The recent failures of the Falcon 9, Antares, Progress, and Soyuz 2 1a expendable launch vehicles are warning signs that NASA management is on a course of certain failure with their expendable Space Launch Vehicle / Orion launch system. As indicated by the FAA chart below, expendable launch vehicles have a "historical" failure percentage of at least 5 percent. Every flight of an expendable launch vehicle is its maiden flight test subject to unknown manufacturing errors. Reusable launch vehicles with proven flight tests are the only option for providing safe and affordable launch service for the twenty-first century._**For NASA management to expect that for their manned Mars mission they can launch up to 16 expendable SLS vehicles in a 26 month period plus operate the expendables trans Earth/Mars vehicles, habitat lander, and crew landers without a vehicle failure is ludicrous. The failure expectation is too great a risk**

Expendable Launch Vehicle Launch Failures Relative to Total Launches
(FAA Data)

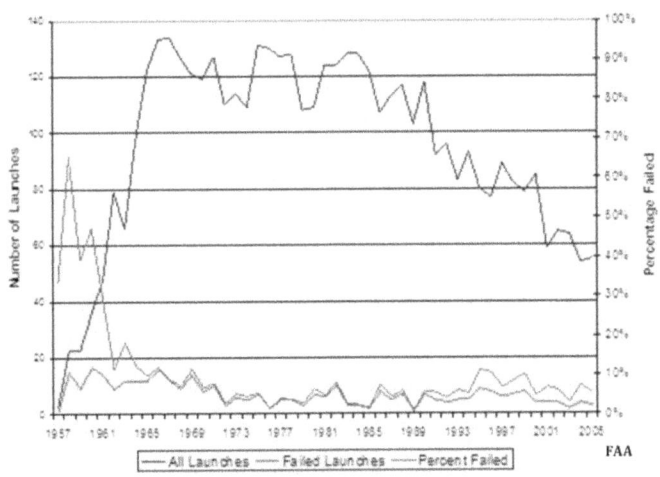

Orion Safety Problem
NASA management promoted the SLS / Orion as being 10 times safer than the shuttle. The ASAP never challenged this flawed assertion until they received a written request recommending that the SLS program be canceled due to unsolvable safety issues. Based on the information presented in this request the ASAP issued the

following statement in their 2014 annual report to Congress (ASAP) expresses concern that the Loss of Crew probability thresholds for them (SLS/Orion) are not significantly safer than the actual historical performance of the Space Shuttle. In other words the Space Shuttles that now reside in museums were as safe as any of NASA's replacement vehicles or the Russian Soyuz launch vehicle and crew module.

In my opinion the NASA advisory council mispresented my warning as evidenced by the following statement made by one member in their meeting report:

"...there is a remarkable difference between mission assurance for commercially developed vehicles and U.S. Government developed vehicles. The U.S. Government failure rates are significantly lower."

What the advisory council member failed to consider was the U.S. Government developed vehicles (Atlas 5 and Delta launchers) number of launches are small when compared to the number of Space Shuttle launches. If continued, the FAA data predicts their failure will be unacceptable.

Author's Email reply to NASA Advisory Council misinterpretation of the safety warning:

Sent: Tuesday, April 19, 2016 1:08 PM
To: NASA Advisory Council, Chairman and Members
From: Don A. Nelson, retired NASA engineer

Subject: Misinterpretation of Space Launch System Safety Warning (December 2015 meeting)

At the NAC December 2015 meeting I presented FAA data (see below) showing that expendable space vehicles like the SLS have an inherent unsolvable safety problem. Expendable space vehicles are exposed to undetected manufacturing errors that can cause catastrophic mishaps. Only reusable space vehicles that have been flight tested can increase flight reliability.

The minutes of the meeting state: "Dr. XXXX commented that there is a remarkable difference between mission assurance for commercially developed vehicles and U.S.

Government developed vehicles. The U.S. Government failure rates are significantly lower. He asked whether those statistics had been considered."

No expendable space vehicle can avoid the possibility of manufacturing errors. The manned Apollo missions suffered three manufacturing error failures; the Apollo 1 fire, the Apollo 13 service module electrical failure, and the Apollo 15 parachute failure. The recent Atlas 5 premature engine shutdown warns that even todays expendables cannot avoid manufacturing errors. As presented, "For NASA management to expect that for their manned Mars mission they can launch up to 16 expendable SLS vehicles in a 26 month period plus operate the expendables trans Earth/Mars vehicles, habitat lander, and crew landers without a vehicle failure is ludicrous." **One manufacturing error can doom both the mission and the crew.**

It has been shoddy NASA management and oversight that has put this nation on this untenable course. Today there are existing technologies that solve the problems of the "reusable" NASA shuttle which would be the baseline for a space base transportation system (www.spacetran21.org). However, NASA management continues on this course of certain failure. **Will the members of the NAC continue to remain silent?**

Don A. Nelson

Retired NASA aerospace engineer

Members of the NASA Advisory Council have remained silent, but this statement in the minutes of their meeting is telling:

"Mr. XXXXX advised that it would be important to have a plan in place for engagement with the next Presidential Administration. Mr. XXXX stated, "If we push the reset button again on SLS and Orion, we are toast."

The Next NASA Administrator

Today in February 2017, NASA has an acting administrator and soon the new president, Donald J. Trump, will name his choice for the new administrator. Will President Trump "drain the swamp" at NASA or will his administration let NASA's human space flight endeavors continue on their path of certain failure? The Trump NASA transition team responsible for advising the president on the problems at NASA has been provided with a draft of *The NASA Letters.*" Draft copies of the manuscript will be provided to the Senate committee that will approve the nomination for NASA administrator. Will those responsible for this decision that could possibly determine mankind's fate finally speak out, or will the deadly silence continue?

Letter to President Trump:

Concerned American Aerospace Engineers
Don A. Nelson, Coordinator
1407 Moller Rd
Alvin Texas
March 15, 2017
The Honorable Donald J. Trump
President of the United States
1600 Pennsylvania Avenue, NW
Washington, DC 20500

Dear Mr. President,

NASA human space flight management has misrepresented the crew safety for the Space Launch System/Orion and the Commercial Crew Program. Both these launch systems programs require crew modules which are potential death traps because they cannot provide crew escape capability for catastrophic failures during entry. Repeated challenges to the NASA Aerospace Safety Advisory Panel that NASA's declaration that the Orion crew module was "ten times safer that the Space Shuttle" finally resulted in the panel's reluctant *admission that the Orion crew module "was not significantly safer" than the decommissioned Space Shuttle.*

The result has been that still viable space shuttles have been consigned to museums and NASA is on a course to develop expendable launch systems that are unaffordable and unsafe. Efforts to get NASA's human space flight management to consider the only viable solution...a commercial space shuttle freighter with crew escape pods has been steadfastly refused. NASA has failed once to develop their unsafe and unaffordable crewed launcher (Constellation) and is in the process of failing again.

Another serious concern is that this nation has no rapid launch capability to respond to the threat of an asteroid/comet impact on Earth. A commercial space shuttle freighter with rapid launch capability will meet this challenge; however NASA space flight management ignores this threat and its solution. NASA human space flight management has us on a course for disaster.

Mr. President, will you appoint a new NASA administrator that will correct NASA's failures of due diligence?

Don A. Nelson
Retired NASA engineer
Coordination: Concerned American Aerospace Engineers

Attn: XXXXX XXXXXX, White House NASA Liaison

Note: On June 30, 2017 President Trump **revived** the National Space Council to coordinate and develop the national space policy. This council will fail like the previous defunct council if they ignore the above warning letter. Their primary goal must be to reduce launch cost and establish a planetary defense system.

Chapter 14 Tomorrow's Challenges

Better get ready...tomorrow is coming!

Planning for tomorrow is always an uncertainty. It's a guess at most, but that guess is more likely to be what actually happens if it's based on information as to what might happen. For many in the human space community their tomorrow vision is planning for humans on Mars. Also the back to the Moon folks are again beating their drums for that to be the next human space goal. Are there other needs that have higher priorities than sending humans to the Moon and/or Mars? Let's consider the following:

Space Debris

NASA Space Debris Map

It's now not uncommon for the space station to be moved to avoid space debris. NASA has reported that there are 20,000 pieces of debris larger than a softball and 500,000 smaller size pieces. When debris impacts a spacecraft at velocities up to 17,500 mph, even the smallest piece can do major damage.

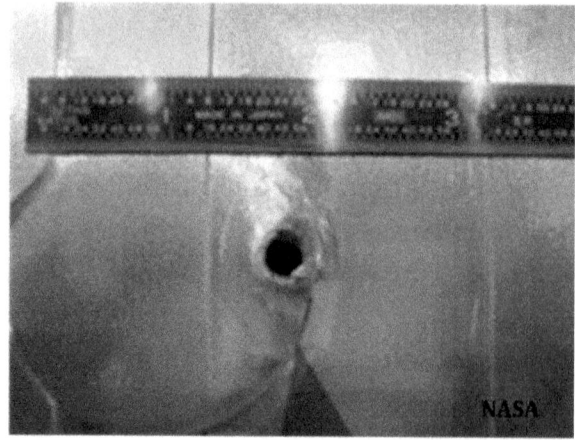

The above picture is of a quarter inch hole in a shuttle radiator that occurred on a 2010 mission. I remember that a hole that size in the orbiter's crew compartment would have released the air to space in about 15 minutes. It would not have been a good day if this small chunk of debris had punctured the crew compartment.

NASA's crew modules are subject to space debris impact which could be fatal. The NASA safety panel has identified this serious safety issue, but since the crew cannot seek shelter in crew escape pods there is no solution. One more data point proving that crew modules are death traps.

Man-Made Space Debris

Another problem is the man-made space debris that falls to earth. Most of it burns up in the atmosphere and if it survives will most likely impact in the ocean. However, most of it is not all of it, and there have been cases where the man-made debris has hit land (see next page):

Propellant tank from Delta II launcher

Somehow this Delta II propellant tank survived entry and impacted near Georgetown, Texas. I remember seeing it in the Johnson Center junk yard, where it stayed for years. Not something you would want dropping out of the sky on you.

Although space debris is a recognized problem within the world's space faring nations, they continue to use expendable launchers, which contribute to space debris, and continue to deploy satellites without deorbit capability. As the problem grows more dire, there is currently little effort to clean up one of our vital resources...near earth space.

Space Environment Effects on Human Anatomy

I think it's a pretty safe conclusion that the long-term effects of the Moon or Mar's gravitational forces on the human body are not going to be good. Also, there is the issue of exposure to radiation levels much higher than we have here on earth. We can solve the gravitation problem during long-duration space travel by creating an artificial gravity, by rotating the crew habitat located at the end of a long structure at a rate to create an equal earth gravity force (centripetal acceleration in circular motion). It's similar to rotating a bucket half-full of water over your head at a rate fast enough so it won't fall out and get you wet (see Mars artificial gravity Transfer Vehicle concept next page).

The crew habitat at left in the picture is insulated to protect the crew from radiation. However, once on the Martian surface the gravitational force drops to about 38 percent that of Earth's and on our Moon it's 16 percent. So in establishing Moon and Mars colonies the long-term gravitational force and radiation on human anatomy again becomes a problem.

Since the human body is about sixty percent water, the body fluids will rise upward in low gravity environments. Some of the side effects will be: bone loss, risks of kidney stones, bone fractures, loss of muscle mass, loss of strength and endurance, diminished cardiac function, blood pressure and volume drops abnormally low, impacts to the neuro-vestibular system, neural sensory loss with failure to maintain balance, vision loss, and anemia (low red blood cell levels). Also, radiation can induce cataracts and cancer. Does this read like the warnings in an advertisement for a prescription drug? The NASA Office of the Inspector General has expressed the following concerns for astronaut's long-term exposure to low or zero gravity and radiation risk in their Oct. 2015 report.

"Space flight is an inherently risky endeavor and NASA has identified 30 human health and performance risks associated with space travel, including Behavioral Health and Performance, Inadequate Food and Nutrition, Space Radiation, and Vision Impairment and Intracranial Pressure. In addition, NASA's current plan to send a crewed mission to the Martian surface by the 2030s will expose astronauts to new and increased hazards."

"Although NASA continues to improve its process for identifying and managing health and human performance risks associated with space flight, we believe that given the

current state of knowledge, the Agency's risk mitigation schedule is optimistic and NASA will not develop countermeasures for many deep space risks until the 2030s, at the earliest."
The ultimate goal of human habitation on the Moon and/or Mars is the establishment of a colony infrastructure. It is logical that human habitation must be feasible and realistic before going to the expense of developing the human transportation systems and infrastructure. Is NASA human space flight management being logical?

Robotic Colonization

The daunting challenges of transporting and protecting humans for any deep space colonization project will require an enormous investment. From the money required point of view, a space base operated by robots is the clear choice. If the project is one for commercial opportunities, robots will most certainly be the primary operators. That path is already being traveled as the following NASA image indicates:

Humanoid robot Robonaut R2 on the Space Station March 2012

While NASA space flight management promotes tomorrow's humans on Mars program, what prevents it from becoming like the Apollo Program, another flag and footprints event? Of course, there is always the standard and valid reply that if mankind is to survive we must be able to go to worlds in deep space. However, what will be the price? Consider the following possibility.

Homo-Mech

Our homo-sapiens body is becoming more robotic as more technology innovations are being used to assist or replace our body functions. It is therefore conceivable that the future deep space traveler may only have the brain of a human and the other parts of the body will be all mechanical...the Homo-Mech space being.

Homo-Mech...mankind's future?

This would simplify the development and operation of crewed long duration space vehicles and facilities. In today's society, the Homo-Mech being would in all probability not be acceptable. Lots of social issues are involved with this concept, like what if the lifetime of the Homo-Mech could be ten or more generations? However, what if spaceship Earth became uninhabitable and the Homo-Mech transition was the only option for survival? In my opinion, there will come a time in mankind's *tomorrow* when the Homo-Mech being will become an issue.

Social Barriers to Space

Climate change, over-population, and world economy are tomorrow's social issues that are and will continue to be issues that indirectly and directly pose barriers to human space endeavors. This old NASA engineer believes that the real issue for climate change is over-population and the continued failure to control population. More and more people require more and more energy generation, creating more heat that warms the earth.

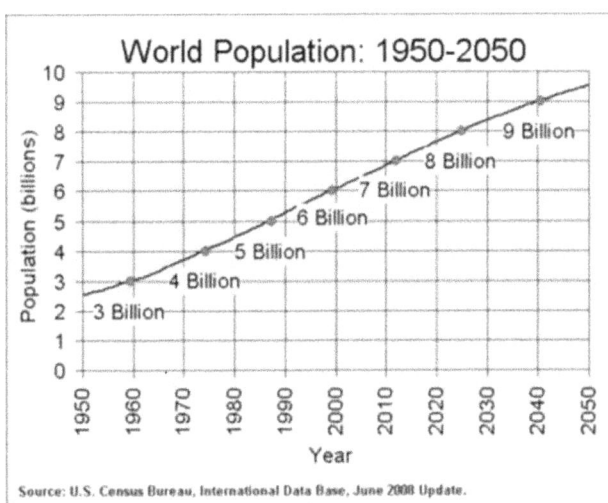

The 7.35 billion people that in 2016 inhabited the planet Earth are more than double that of fifty years ago. As the above population chart indicates the world population growth has shown no indication of slowing down. When the population increases, the need for more manufactured goods and food production increases, all of which increases the world's heating factors. Add to this the world's government expenses needed to provide protection and care for this growing population, and funding for space endeavors will become more strained.

International Cooperation for Space Infrastructure

The issues dividing the earth's societies are many. Religion, method of government, and economic inequalities are just a few. What we need to bring us together is a common bond. I believe that common bond is a space infrastructure. The International Space Station has been the first step in securing that bond of cooperation in a common cause.

The next step should be the creation of an international group dedicated to the protection of earth from deep space threats and eventual deep space colonization. NO NATION SHOULD BE EXCLUDED! We must learn to live together or we will most certainly all die together.

The Deep Space Threat is Real

In the Introduction of this book I describe a fictitious event where the asteroid *Bad Day* was on an impact trajectory to earth. This old NASA engineer has an uneasy feeling about the threat to Earth from an asteroid or comet impact. It's the same kind of feeling I had about the Space Shuttle Columbia disaster…just too many warning signs of a coming disaster. One warning sign was the visual sighting of the two asteroids that impacted the planet Jupiter in 2009 and 2012. Though these asteroids posed no threat to Earth, they are sobering recent events that remind us that the threat still exists. Other warning signs are the near-earth asteroids and comets misses that have been recorded recently. NASA's Near Earth Object Program Office so far has recorded ten recent close approaches of asteroids and comets and has eighty upcoming close approaches on their books.

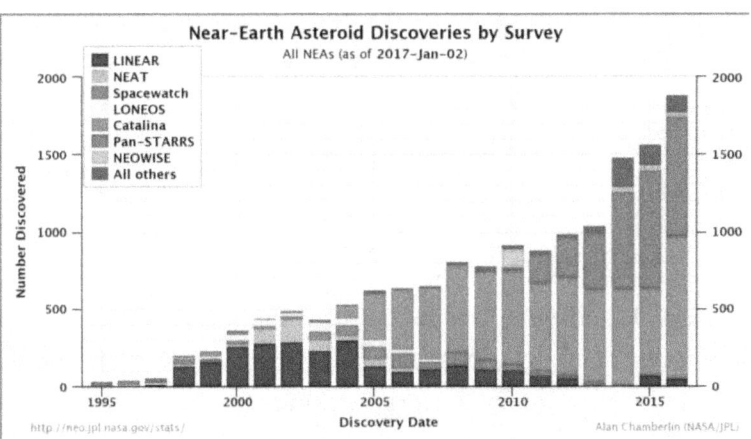

What is very concerning about asteroids and comets is, first, some are discovered too late to take any action if they are on an earth impact course and, second, is the uncertainty in their size, composition and direction. This means that we will in all probability have little time to determine if an asteroid or comet is a threat to earth and even less time to destroy or change its path. Consider the effects of the 1908 impact in Russia and what the results would have been if it had hit in New York City.

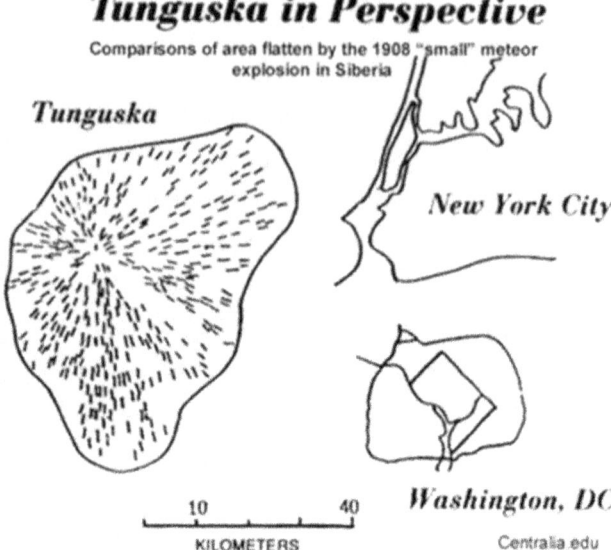

Tunguska in Perspective

Comparisons of area flatten by the 1908 "small" meteor explosion in Siberia

Tunguska

New York City

10   40   *Washington, DC*

KILOMETERS   Centralia.edu

We must have a rapid response launch system to analyze and counter this very real threat to earth, and only the CSS freighter can meet that requirement.

Which Path to Tomorrow?

In the judgment of this old NASA engineer, *tomorrow's* path to take humans to Mars or back to the Moon is not our highest priority. First, we must get our launch cost under control, and the only feasible and realistic way to accomplish that is the Commercial Space Shuttle freighter. Second, to keep operation cost down we must have a near earth and deep space transportation system that is space based and commercially operated. Third, we need to get NASA out of the launch operation business and back to research and development of space systems.

Note: On March 2, 2017 a "small" asteroid identified by NASA's Near Earth Object Program Office as Asteroid 2017 EA came within 9000 miles of impacting on our planet. According to one space authority if the asteroid was an iron meteoroid and entered the earth's atmosphere **there could**

have been an explosion equal to the atomic bomb that destroyed Hiroshima.

Tick, tock, tick, tock… is mankind running out of time?

SPEAK OUT NOW…*TOMORROW* MAY BE TOO LATE. TELL THE PRESIDENT AND CONGRESS, WE NEED THE COMMERCIAL SPACE SHUTTLE FREIGHTER AND A PLANETARY DEFENSE SYSTEM!

Note: On October 12, 2017 asteroid 2012 TC4 is predicted to miss space ship Earth by at least 4200 miles…if NASA made a miscalculation you may not be reading this book!

www.ingramcontent.com/pod-product-compliance
Lightning Source LLC
Chambersburg PA
CBHW051638170526
45167CB00001B/242